David Cromwell

Always be kind
and caring.

Janet Williams Roland
12/2011

HARSH COUNTRY, HARD TIMES

Number Thirteen:
CLAYTON WHEAT WILLIAMS TEXAS LIFE SERIES

HARSH COUNTRY, HARD TIMES

*Clayton Wheat Williams and the
Transformation of the Trans-Pecos*

**JANET WILLIAMS POLLARD
& LOUIS GWIN**

Texas A&M University Press
College Station

This paper meets the requirements of ANSI/NISO z39.48-1992 (Permanence of Paper).
Binding materials have been chosen for durability.
∞ ♻

Library of Congress Cataloging-in-Publication Data

Pollard, Janet Williams.
    Harsh country, hard times : Clayton Wheat Williams and the transformation of the
Trans-Pecos / Janet Williams Pollard and Louis Gwin.—1st ed.
        p. cm.—(Clayton Wheat Williams Texas life series ; no. 13)
    Includes bibliographical references and index.
    ISBN-13: 978-1-60344-283-1 (cloth : alk. paper)
    ISBN-10: 1-60344-283-9 (cloth : alk. paper)
    ISBN-13: 978-1-60344-479-8 (e-book)
    ISBN-10: 1-60344-479-3 (e-book)
    1. Williams, Clayton. 2. United States. Army—Officers—Biography.
3. World War, 1914–1918—Campaigns—France. 4. Ranchers—Texas—Biography.
5. Historians—Texas—Biography. 6. Petroleum industry and trade—Texas—History.
7. Statesmen—Texas—Biography. 8. Fort Stockton (Tex.)—Biography. 9. Trans-
Pecos (Tex. and N.M.)—History—20th century. 10. Texas—History—20th
century. I. Gwin, Louis. II. Title. III. Series: Clayton Wheat Williams Texas life
series ; no. 13.
    F392.P3P65 2011
    976.4'061092—dc22
    [B]
                                                                          2011007988

Contents

Acknowledgments

This collaboration had its genesis in 2007, when Janet Pollard met Bill Krumpack while traveling with a group of Austin opera buffs in Italy. When Janet told Bill that she needed someone to help her organize a book about her father's life, he mentioned Louis Gwin, a retired associate professor of communications who had considerable experience as a writer. After several telephone conversations and a meeting in Midland that summer, the project was born.

This book has benefited from the technical expertise of a number of people. The authors are particularly grateful to Jim Bradshaw, archivist of the Nita Stewart Haley Memorial Library and J. Evetts Haley History Center in Midland, where the papers of both O. W. Williams and Clayton Wheat Williams are housed. Jim spent many hours tracking down letters and other documents that were essential to telling this story.

Dennis Trombatore, head librarian of the Walter Geology Library at the University of Texas at Austin, was an invaluable guide to books, periodicals, and other materials documenting the development of the West Texas oil industry and drilling technology. Professor William L. Fisher, the Leonidas T. Barrow Centennial Chair in Mineral Resources in the Department of Geological Sciences at UT–Austin, gave freely of his time to review chapters covering early technological developments in the Texas oil industry. Howard Parker, now retired in Austin after a successful career as a West Texas oilman, provided many interesting details about oil exploration techniques. Debra Whitfield, librarian for the Fort Stockton Public Library, was a wonderful resource for information about the history and growth of Fort Stockton. And while there are many other individuals who contributed to this effort, we would be remiss without thanking the collective staffs of the Dolph Briscoe Center for American History at UT–Austin and the Permian Basin Petroleum Museum in Midland for guiding the authors to a number of important sources of information about the life of Clayton Wheat Williams and the Trans-Pecos region of Texas. Of course, any errors of fact in this book are solely the responsibility of the authors.

On a personal note, the authors wish to acknowledge people who encouraged and supported them throughout the work of putting this book together. Louis Gwin thanks his wife Rachel Parker-Gwin, PhD, who read the manuscript several times and contributed her intellectual insights as a sociologist and her considerable proofreading skills, as well as her love and encouragement.

For Janet Pollard, this book is the culmination of a promise made to her father more than two decades ago. Bringing the book to fruition was a long and difficult task, and chief among those who supported her dream along the way was her beloved husband Bob, who encouraged her and loved her for forty-eight years; and her strong sons, Scott, Clay, Graham, and Adam, who have helped her and taught her. Thanks also go to their loving wives, Debbie, Jeanne, Chris, and Hollyn, and to Janet's grandchildren Ashlee, Meredith, Stephanie, Clayton, Mary Claire, Addison, Andy Bob, and Jack.

Janet's grandfather, O. W. (Judgie) Williams, has her gratitude for his dedication to learning and his great sense of adventure, and her grandmother, Sallie Wheat, for the gift of music, which she passed down. Janet's parents gave her a joyful childhood. Janet is particularly grateful to her father for asking her to write this book and for believing that she could do it. Her brother Claytie deserves mention for all of the wonderful times they have spent together, and his wife Modesta, for loving them all.

She thanks her Williams cousins—Mari Helen Schultz, Sara Garnett, David Walker, Harriet Hern, Susan Poole, Ann Kirk, Bill Hamilton, Marsha Williams—and her nieces and nephews—Kelvie Cleverdon, Allyson Groner, Clayton Williams, Jeff Williams, Chim Welborn, and Shelly Pollard Forte—as well as her Graham cousins—Lee Graham, Nancy Graham McSpadden, and Joe Graham—and Dick and Tina Pollard.

To her friends who stuck with her—Nancy Beal, Jane Sibley, Barbara Bross, Mary Carney, Pat Black, Peggy Schafer, Mary Rasmussen, Shirley McDonnold, Sharon Floyd, Karen Williamson, Betty Isaac, Carolyn Roden, Billie and Jim Ross, Dulcie Ligon Boykin, Jane Adams Garlitz, Kay Salyer, Dorothy Blackwell, Francis Levine, Lisa DiLeo, and Teresa Lozano Long—the author offers her sincerest love and appreciation. To Frances Stapp, who helped her move 600 boxes of Williams family papers to the Haley Memorial Library and History Center in one trip, and to J. Evetts Haley and Jim Bradshaw, who so carefully archived the collection, she owes a huge debt of gratitude, as she does to Bill Krumpack for introducing her to Louis Gwin and his wife Rachel. And Ed Todd and Georgia Temple provided continued encouragement.

Finally, she credits the First Baptist Church of Midland and the prayer groups and Sunday School classes that have helped sustain her for the past nineteen years.

Prologue

The maroon and white Saberliner, painted in the colors of Texas A&M University, caught the gleam of the bright July sunlight as it lifted off the runway of the Midland airport and headed for the small West Texas town of Fort Stockton. The people on board focused their attention on the handsome elderly gentleman who was wearing an oxygen mask—Clayton Wheat Williams. Though he was seriously ill, the mask did not detract from his dignity and courage as he made the sad journey home.

With these sentences, I began years ago to write the story of my father's life, having promised to complete the autobiography he had begun in his final year. Clayton Williams's story is one of rugged individualism and risk-taking in a period of transformation for the Trans-Pecos region of Texas. It begins in Fort Stockton at the turn of the twentieth century, continues through the development of Texas as a major oil and gas producer between the two world wars, and ends in the West Texas landscape of small towns, large ranches, water wells, and oil-pumping units. My father, like his father before him, helped tame a harsh land and extracted from it the resources of oil, gas, and water. And also like his father, he never lost his love for the land's rich heritage, spending his last years documenting the colorful history of the region and its people.

Somewhere below the wings of his son's airplane were the Pecos River and the Grandfalls ranch where Clayton had lived as an infant, a place that represented many years of hardship for his mother and father but a place also filled with cherished memories.

Oscar Waldo Williams (O. W.), Clayton's father, had come to the pioneer hamlet of Fort Stockton in 1884 with his young family, his Harvard law degree, and his surveying instruments. By his death in 1946, O. W. was one of Fort Stockton's leading citizens and had earned a reputation for integrity and honesty as a county judge and lawyer. In the early days, while trying to scratch out a living on the ranch north of town, he had made ends meet by accepting surveying assignments in the vast, uncharted landscape that surrounded Fort Stockton.

It was on these trips into the wilderness that O. W. honed his skills as a keen observer of the land and its inhabitants, both two-legged and four-legged, and he later recorded these observations in stories published in pamphlets for his children and grandchildren. Clayton and his older brother Waldo accompanied their father on some of these trips while boys. Years later, Clayton applied early lessons from his father about "reading" the landscape to a successful career as an oilman and rancher. Like O. W., he explored the harsh land, though his efforts initially focused on what its surface features could tell him about what was hidden underneath.

In less than twenty-five minutes the airplane was touching down in Fort Stockton. There to greet Clayton and his wife and family was a large group of people both young and old, his friends from every walk of life. "Welcome Home Clayt & Chic" and "We Love You" were painted on signs held by friends. Signs and balloons were everywhere. Clayton was put in a limousine and mariachis serenaded him while friends came by to shake his hand and wish him well. All around his car family and friends cried and hugged as they acknowledged their beloved Clayton.

Throughout his life, Clayton was a risk taker, tackling tasks that would have prompted another to back off or go around. He developed toughness as a youngster and as a cadet at Texas A&M—toughness that would serve him well both in the war to come and later in the rough-and-tumble life of the oilfields. Seeing that his country needed officers as it entered World War I, he volunteered before the draft was instituted and served with distinction in France and Flanders. His letters home, inspired by his father's urging to record his experiences in writing, provide an unusually nuanced picture of what life was like for an American officer in France during the war.

Returning to Texas after the war, Clayton recognized the opportunities in the developing Texas oilfields and taught himself to become a first-rate petroleum geologist—so good, in fact, that he picked the site of what would become in 1928 the deepest producing oil well in the world. He and his brother Waldo also mapped the structure of what later became the Fort Stockton oil and gas field. From 1924 to 1980, Clayton joined with his father and later his son to hammer out a successful career in the boom-and-bust cycles of the West Texas oil industry. Indicative of his achievements was his election to the Permian Basin Hall of Fame after his death in 1983.

But apart from his successful business career, many people in Fort Stockton remember Clayton as a particularly civic-minded man who served as a Pecos County commissioner for fourteen years and as an officer in many social and civic organizations. Others know him because of his love of the region's history, having read his stories in the local newspaper, heard him speak to countless groups, or read his books. And some also remember him for what they believe was his role in drying up the town's famous Comanche Springs by pumping water feeding the spring's aquifer to irrigate his farm west of town.

When we arrived home, I asked my Dad if the trip and the reception at the airport had been too much for him. He replied with a warm smile, "When I came home from World War I, there was no one to meet me as I stepped off the train. Today makes up for that."

What makes Clayton Williams's life story worth telling? As noted above, he was a member of a generation that returned from World War I to survive the Great Depression and master a new industry that sparked the development of West Texas. His life spanned a dynamic period in Texas history when automobiles replaced horse-drawn wagons, electricity replaced steam power in the oilfields, and barren and virtually worthless ranchland became valuable for the oil, gas, and water under its surface. Finally, through his books and other writings, Clayton documented in rich detail the history of an unforgiving land and the people who confronted its challenges.

This book is not intended to be an objective, scholarly analysis of Clayton's life, although the authors have tried to place the events of his life within a historical context. Instead, Louis and I have tried to write a serious biography based on hundreds of letters and personal papers left by Clayton and his father, on the reminiscences of family and friends, and on interviews that Clayton gave in his final years to preserve his memories of what life was like in the Trans-Pecos during his time. Our goal is to give readers an honest picture of a man who both shaped and was shaped by the country in which he was born, lived, and died.

In the end, however, this book is the result of a daughter's promise to and love for her father. While life's ups and downs—the raising of four sons and the untimely death of my beloved husband Bob—have slowed the fulfillment of that promise, I have worked on this book over the years in numerous ways. I collected and organized my father's many letters and personal papers. I visited the towns and cities in France where he was stationed during World War I. I traveled throughout Texas and New Mexico to walk the same ground that my father and grandfather walked. And I listened to and recorded stories told by relatives and friends about my father.

It is at my father's special request that I write the story of his life. He left many handwritten pages and legal files bulging with old letters and stories that I found at his home and office. During his final two months my family taped as many of his stories as possible, but I wish that we could have recorded more, for many questions remain unanswered and there are so many more stories that I wish I had recorded. That is the way life goes.

Acknowledging that I am not a writer, but with the assistance and editorial support of Louis and with great love in my heart for my father, I attempt to tell his story.

Janet Williams Pollard
Midland, Texas
August 2009

HARSH COUNTRY, HARD TIMES

CHAPTER I

"Paradise for Men and Dogs and Hell for Women and Horses"

Interstates 10 and 20 cut through the Trans-Pecos region of Texas like arrows before joining to become Interstate 10 about 150 miles east of El Paso. Most people traveling these highways probably wouldn't recognize the term *Trans-Pecos* and simply call the passing landscape *West Texas,* but the Trans-Pecos is a distinct region, bordered on the east by the Pecos River, on the south and west by the Rio Grande, and on the north by the 32nd parallel that forms the boundary with New Mexico.[1] The Midland-Odessa area of more than 250,000 residents is the largest population center in the region and serves the oil and natural gas industry, but most of the towns are small rural communities and some are merely shells of buildings without people, having boomed and burst with the cycles of oil and gas development since the 1920s.[2]

The cars and trucks speeding toward either El Paso or San Antonio along Interstate 10 pass by three exits to the town of Fort Stockton, a community of 7,600 located about seventy miles east of where I-10 and I-20 meet.[3] Fort Stockton and the area to its south are located in the Chihuahuan Desert, a series of hills, mesas, and peaks covered with short grasses, bushes, and a few trees. Farther to the south and the west near Alpine and Marathon are the Davis Mountains, foothills of the Rockies.

Water is life in this arid land, and the scattered springs that once dotted the area were critical to the early settlers. Fort Stockton grew up around one such spring, and it was a source of clean water for those who passed through on their way west and for those who settled the small farms irrigated by Comanche Springs and the waters of the Pecos River to the north. For the newcomer, the desert around the town was foreboding—"paradise for men and dogs and hell for women and horses."[4] But for the people who lived there, the clean, dry air magnified the sun, and the huge sky became a canvas on which sunrises, sunsets, and the occasional thunderstorm painted brilliant colors of gold, pink, purple, and lavender.

Fort Stockton was a farm and ranch community when Clayton Wheat Williams entered the world on a Monday in April 1895. The fourth of five children born to Oscar Waldo (O. W.) Williams and his wife Sallie Wheat, Clayton arrived at about forty minutes past 4 P.M. in the southernmost house of what was known as "officers' row" in the old military garrison. It had been built by the U.S. Army in 1859 to protect settlers and the mail service from Indian attacks. O. W., who served as Pecos County judge while at the same time trying to make a go of a ranching operation some twenty-eight miles north of town on the south bank of the Pecos River, had leased the house as a base from which to conduct his business in town.

Clayton's older brother, Oscar Waldo II, or Waldo, for short, was born in 1883, and sister Mary Ermine was born two years later, both at their mother's family home in Dallas. Susan Kathryn, who would be known to the family as Kathryn, was born at the ranch in 1892, and Clayton's younger brother, Jesse Caleb (J. C.), with whom he shared a birth date, was born four years after Clayton in the officers' house.

Before O. W. settled in Fort Stockton with his family in 1884, he spent several years surveying the desolate plains of West Texas, returning periodically to the East to seek buyers for the lands at a price of thirty cents per acre. When the Texas legislature passed a law in 1879 to reserve unappropriated lands from private sale, O. W. turned to silver mining in frontier New Mexico, where he faced dangers from both Indians and his fellow miners. During his years in the wild lands of Texas and New Mexico, O. W. recorded his observations, experiences, and adventures, which he later set down in a series of letters to his children.[5] A keen observer of daily life and a dedicated recorder of events, O. W. tried to instill in his children, and especially his sons, a belief that the past would be remembered only if it were accurately recorded. He summarized his philosophy in a 1918 letter to J. C., who was away at college while Clayton was soldiering in France. After praising Clayton's letters home for their description and detail, O. W. emphasized the importance of first observing events and then thoughtfully recording what was seen. "These 2 things," he wrote, "are of the greatest importance in life—to see things & to think over them. I find men—going side by side—one noticing keenly—the other hardly at all—one considering the meaning of the things he sees—the other passing them out of mind with out a thought, there can be no question which one of the two gets the good out of life."[6]

Clayton and J. C. took the advice to heart. Clayton's letters from France provided a nuanced view of what life was like for a young officer, while J. C. sent home many letters describing his experiences in China while working for a subsidiary of Texaco, and these were later published by the Williams family as a series of pamphlets. Clayton's grandson, Adam Pollard, recalls that his grandfather always encouraged him to write letters because letter writing had become a dying art. "When I was a boy, I would write him from camp and could not wait for his return letter. He could describe things and events as if you were sitting right there with him. I still do a lot of my negotiations via handwritten letters. It adds that personal touch my grandfather always spoke of."[7]

Like his father, Clayton believed in writing as a means of recording history, and in later life he completed a series of writing and editing projects that explored his interests in Texas history and pioneer life.[8] Both men also had a keen sense of personal dignity and a measure of reserve. As historian C. L. Sonnichsen notes in the introduction to *Pioneer Surveyor, Frontier Lawyer: The Personal Narrative of O. W. Williams, 1877–1902,* O. W.'s published writings and letters, though accurate and observant, dealt with facts but very few feelings.[9] Clayton's writing took a similar path. Both were tough and fearless, and neither was afraid to solve a problem in the rough-and-tumble of West Texas life with his fists or the threatened use of a gun. But the two differed in other ways. While O. W. looked harshly on those he believed took part in frivolous and time-wasting activities, Clayton was not averse to dancing, having an alcoholic drink now and then, and telling the occasional joke or story.

While Clayton shared many of his father's traits, particularly O. W.'s passion for observing and recording, he was also very much his own man. A risk taker at heart, Clayton was frequently at odds with his father's motto of "safety first" when it came to financial matters, and the judge viewed with skepticism his son's rush to plunge headlong into what O. W. believed were risky financial ventures, particularly in the fledgling Texas oil and gas industry of the 1920s. Years later, Clayton would give the same cautious advice to his son, nicknamed Claytie, and like his father before him, Claytie ignored it.[10]

"I have witnessed my grandfather scolding Claytie about taking foolish risks with his money in the oil business," said grandson Scott Pollard. "It was fascinating to learn that my great-grandfather scolded my grandfather in the same way. I can only imagine what my great-grandfather's father thought of his son. Each of the sons ignored their fathers and took on risks—risks the fathers knew well and hoped they could avoid for their sons. Each son had to learn for himself. They all had success, but I think at its root it was not money they were after but the adventure."[11]

Later in life, O. W. changed his mind about the oil business and became an active participant with his sons, using his legal expertise and knowledge of the countryside to purchase and execute oil leases for himself, his family, and others. He also helped his sons finance several speculative oil drilling projects in Pecos County. Clayton as well joined Claytie in several oilfield projects and often wished that his father could have seen the accomplishments of his children and grandchildren.

Scott Pollard believes it was O. W.'s emphasis on the value of education that enabled the Williams family to take advantage of the new technological age that the discovery of oil would bring. "Swept up in the enthusiasm for the age of petroleum and converted by the knowledge of the workings of the oil patch, my great-grandfather and his sons became oil people. Because of their talents, the oil patch enticed them out of their West Texas homes and cast them about the state, the nation and the world."[12]

A Family Tradition

In attempting to make a living in the desolate and rugged West Texas countryside, both O. W. and Clayton followed in the footsteps of their ancestors, men and women who had a high sense of adventure and were indifferent to the dangers in a new country. To understand the character of the man who was Clayton Wheat Williams, it is necessary to understand the ancestral tree from which he was descended. Like his father, Clayton was fiercely proud of his forebears, who settled in Virginia from England as early as 1639, and both father and son took pleasure in the fact that the men and women who preceded them participated in some of the great moments of American history.[13]

For example, O. W.'s great-grandfather on his maternal side, John Colyer (later Collier),[14] served in the French and Indian War as a drummer boy for British Gen. Edward Braddock and was present during Braddock's great defeat in 1755 at the Battle of the Monongahela in present-day Pennsylvania. According to the family genealogy compiled by O. W.'s sister, Jesse Williams Hart, Colyer was about thirteen in 1755, and his service in the war "would have been in a minor capacity."[15] Colyer also served in the Revolutionary War, enlisting in either 1776 or 1777 in the Virginia Continental line. According to Hart, the old soldier told his children and grandchildren that he was at Valley Forge in the winter of 1777–78 and told stories about the hardships suffered by soldiers during those harsh months.[16]

Department of the Interior records reproduced by Hart show that Colyer also served as a sergeant in Gen. Arthur St. Clair's army in the 1791 war against the Ohio Indian tribes in the Northwest Territory. During this war, in a battle near the headwaters of the Wabash River in northwestern Ohio, the U.S. Army suffered the greatest defeat against Indian forces in its history. According to Hart, Colyer was wounded in the battle and coughed up blood as he retreated, which allowed the Indians to follow his trail in the snow. When he realized he was leaving a trail, Hart said, Colyer "opened the front of his heavy hunting shirt and spat the blood into it." She said Colyer hid in the hollow of a dead tree and "heard the conversation of Indians" as they pursued the survivors.[17] Successfully evading his pursuers, he eventually made his way to Fort Jefferson, about six miles south of present-day Greenville, Ohio.

Colyer was lucky to have survived the battle. Under the leadership of Little Turtle, war chief of the Miami Indians, some 1,400 warriors from several different tribes routed the U.S. troops in a three-hour battle on November 4, resulting in 634 U.S. dead out of 1,400 present. Indian losses were estimated at between 100 and 120. Hart said that when Colyer died in Kentucky in 1826, a dozen bullets that had been extracted from his body over the years were placed in the coffin with him.[18]

Richard Gott Williams and the Trail West

Ensign Jesse Williams, O. W. Williams's paternal great-grandfather, also served in the Revolutionary War, and his son, Richard Gott Williams, was one of the first merchants to head west on the Santa Fe Trail.[19] The story of Richard Gott Williams's journey was recorded in detail by his son, Jesse Caleb Williams (O. W.'s father), and O. W. and Clayton also retold the story in various writings. Although there are some features of the story that cannot be independently confirmed, this saga of settling the pioneer west embodied elements that have since shaped the Williams family spirit—the taking of great risks, both physical and financial, in search of a better life while combating a harsh natural environment that destroyed lives but also created great fortunes.

In 1826 Richard Gott Williams traveled west on the Santa Fe Trail with several wagons. This was a mere three years after George Sibley began his historic survey of the trail.[20] Williams was born in 1786 on his family's farm in southeastern Culpepper County, Virginia, near the town of Fredericksburg. A saddle and harness maker by trade, he served his apprenticeship in Richmond, Virginia, and developed a reputation as a sound craftsman. In 1808 he moved his business to Richmond, Kentucky, where he prospered. He married Catherine Holder, the orphaned daughter of Capt. John Holder and Frances Callaway Holder of historic Boonesborough, Kentucky, in 1812.[21]

Col. Richard Callaway, Holder's father-in-law and O. W.'s great-great-grandfather, was one of thirty men selected by Daniel Boone in 1775 to clear Boone's Trace, later renamed the Wilderness Trail.[22] Holder, one of the early Boonesborough settlers, married Callaway's daughter Frances in 1776. The marriage occurred a few weeks after Frances, her sister Betsy, and Jemima Boone, Daniel's daughter, were captured and held by Indians for two days. The children were rescued by Boone and a group of men that included

Richard Gott Williams (1786–1876), Clayton Williams's great-grandfather, who headed west across the Santa Fe Trail in 1826, perhaps accompanied by young Kit Carson.

Holder. Holder's date of death is uncertain but was probably the winter of 1797–98, and Frances died in 1803. Their daughter, Catherine, was brought up by relatives.[23]

Richard Gott Williams obtained a contract during the war of 1812 to procure lead for the Kentucky military from mines located west of St. Louis. Excited by news of the trade opportunities on the Santa Fe Trail, Williams outfitted wagons and transported them to Westport, Missouri, for the journey west.[24] Jesse Caleb Williams did not provide any details of the trip on the Santa Fe Trail, but O. W. added an intriguing note to the story. It is historical fact that sixteen-year-old Kit Carson made his first trip west on the Santa Fe Trail in 1826, the same year as Richard Gott Williams. In a 1926 letter to J. C., O. W. said that Carson traveled in the Williams's caravan and that Carson and the men from Richmond, Kentucky, shared some common acquaintances. C. L. Sonnichsen, in his introduction to *Pioneer Surveyor, Frontier Lawyer,* said that in 1826 Richard Gott Williams outfitted "a caravan of twelve prairie schooners" and traveled west across the trail, but made no mention of Carson's presence on the trip.[25]

One of the party, Andrew Broadus, was seriously injured in a rifle accident during the trip. Carson relates the story in his autobiography, and it has been told by others.[26] For O. W., the story of Broadus's accident was the key evidence connecting Carson to his great-grandfather in time and place on the Santa Fe Trail. He cited a conversation he had with ninety-three-year-old Samuel Parke during a trip to Richmond in 1923. Parke claimed his uncle, James Parke, accompanied Richard Gott Williams on the caravan west in 1826 and that Broadus's accident had occurred on the same trip.[27]

One historical record shows that a company of traders, size unknown, left Fort Osage on the Missouri River in August 1826 and reached Santa Fe in November. Included in that company were Carson, Broadus, James Collins, Elisha Stanley, Soloman Houck, Edwin M. Ryland, James Felding, Thomas Talbot, William Wolfskill, and possibly George Yount. In his autobiography, Carson did not name any of the men with whom he traveled except Broadus, making it likely that this was the group of traders that Carson joined. However, neither James Parke nor Richard Gott Williams was named in this record.[28]

Richard Gott Williams's trip to Santa Fe was made without further incident. Jesse Caleb Williams said his father took the caravan to Taos, then to northern Mexico before disposing of his remaining goods. With his wagon train, bills of exchange, money, and a herd of 1,500 horses and mules, 500 of which he owned personally, Williams traveled back over the trail from Santa Fe. According to varying accounts told by Jesse Caleb, O. W., and Clayton, the party was attacked by Indians in either New Mexico or Kansas while returning east, and most of the horses, mules, and other possessions were stolen. O. W. said that his relatives in Kentucky put the loss at more than $100,000, while Clayton, writing in *Never Again,* said the loss was $112,000 in currency and 800 horses and mules.[29]

According to Jesse Caleb Williams, Richard Gott Williams petitioned Secretary of State Henry Clay for reimbursement, claiming that he was attacked by "Indians of the United States" and that the government was liable for his losses. Clay responded that because the border between Mexico and the United States was in dispute, the location of the attack could have been on Mexican land and that Mexico was liable for Williams's loss. In his introduction to *Pioneer Surveyor, Frontier Lawyer,* Sonnichsen said that Richard Gott Williams filed a complaint in a Mexican court in 1826 alleging that a thief had stolen some of his goods, but apparently Williams did not recover any of his loss. He was, however, able to sell some of his property in Kentucky and buy an undeveloped plantation in Rockcastle County that he worked from 1829 to 1847. He moved to Mount Vernon, was appointed postmaster, and revived his saddle and harness making business before he died in 1876 at the age of ninety.[30]

The historical record shows that after a year trading goods in Mexico, the traders who accompanied Kit Carson west, including Collins, Stanley, Houck, Ryland, Felding, Talbot, and Wolfskill, were in a caravan on the Santa Fe Trail traveling back to Missouri when they were attacked by Indians. They described the attack in a letter to the United States Congress in which they petitioned for financial relief for their losses. According to the letter, the attack occurred on October 12, 1827, while the caravan was camped on the north side of the Arkansas River in present-day Edwards County, Kansas. The Indians were able to drive away 166 horses and mules, and after giving chase, the traders were able to recapture sixty-six of the animals. Neither Richard Gott Williams nor James Parke was mentioned in the petition to Congress.[31]

O. W. Williams

Clayton's father, O. W. Williams, was born in 1853 in Mount Vernon, Kentucky. He was the son of Jesse Caleb, who ran a general store, and Mary A. (Collier) Williams, the granddaughter of the Revolutionary War soldier John Colyer.[32] O. W. was the second of seven children. An older sister, Catharine, was born in 1851 and died a year later.

The other siblings were William David Williams (born in 1855), Josiah Joplin Williams (1858), Edward Everett Williams (1860), Anna Susan Williams (1862), and Jesse Williams (1865).[33]

Clayton's grandparents and their children at the family home in Carthage, Illinois, taken about 1900. From left, front row, Clayton's uncle William David Williams, grandfather Jesse Caleb Williams, grandmother Mary Ann Collier Williams, and father Oscar Waldo Williams. Back row, aunt Anna Susan Williams, uncle Josiah Joplin Williams, and aunt Jesse Williams Hart.

When O. W. was four, the family moved to Carthage, Illinois, where Jesse Caleb established another general store and served in both the Illinois General Assembly and the State Senate.[34] The family home in Carthage was located on a large tract of land at the edge of the prairie, and O. W. fondly recalled gathering nuts in the nearby woods as a child and watching his mother at her spinning wheel. "Now and then in the quiet hours of wakeful nights I can hear its peculiar whirr and hum as my mother sat before it. So long as it was running there was no hope that we could get her to tell us what we so greatly enjoyed—stories of the Revolutionary War and tales from the Bible and the classics; so we would turn away to play."[35] One of O. W.'s earliest memories from Carthage was seeing Abraham Lincoln make an appearance on the courthouse square in 1858. Later, O. W.'s father would, like Lincoln, become an actor on the Illinois political stage; he would serve in both the Illinois General Assembly and the State Senate.[36]

After six years of public education and some tutoring in Latin at home, O. W. was taken by an uncle to Liberty, Missouri, for a year of schooling at

*The Williams family home in Carthage, Illinois, where O. W. Williams sent his older children
from Fort Stockton to live with his parents and receive most of their education. Clayton
spent one year in Carthage before returning home.*

William Jewell College, and then he attended Christian University (now
Culver-Stockton College) in Canton, Missouri, not far from his home in
Carthage. Feeling that his son was devoting too much time to his Carthage
friends and not enough to his studies, Jesse Caleb relocated O. W. to Bethany College in the northern panhandle of West Virginia. As a devout member of the Christian Church (Disciples of Christ), Jesse chose Bethany not
only because it was in a rural area, far from the temptations of Carthage, but
also because it had been founded by Alexander Campbell, who also founded
the denomination. There O. W. spent nine months, and he was awarded a
degree in civil engineering in 1871.

He worked as a civil engineer in railroad construction near Fairmont,
Missouri, and as deputy county clerk of Hancock County, Illinois, before
deciding to undertake a legal education. He enrolled at Harvard in 1873, but
after two years of study failed his examinations. He tried various jobs for a
year, including managing a store in Clayton, Illinois, and then returned to
Harvard in May of 1876, where this time he passed his examinations and, at
the age of twenty-three, took his law degree.

O. W. briefly practiced law in Chicago but developed a lung condition
that his doctor believed would improve in a drier climate. He arrived in
Dallas in 1877, concluded that the town had too many lawyers, and took a
job as a surveyor of public lands. He made three expeditions to the Texas

Panhandle and West Texas and years later recorded his observations and experiences in a series of printed pamphlets that he distributed to his children
and grandchildren. In the summer of 1878, while surveying near the site of
the town of Plainview, O. W. saw one of the last great herds of bison remaining in Texas. He estimated their number at between 10,000 and 40,000
animals.[37]

When the law on selling unappropriated lands changed in 1879, O. W.
headed for New Mexico to prospect for silver, traveling that summer by
horseback from Dallas to the Cerrillos hills with five men. Here, in a
mining camp about twenty miles south of Santa Fe, he again recorded his
observations of the rugged country and the lives of the prospectors and
others who tried to make a living there. He returned to Dallas in the fall
of 1879 and the following spring traveled to Silver City, having heard positive reports about the silver mining in that part of southwest New Mexico.
To make ends meet while waiting for his big strike, O. W. held positions
in Silver City as deputy postmaster and as Grant County deputy district
clerk. He also became part owner of the Oasis silver mine in Lordsburg,
New Mexico.[38]

One of the men who rode with O. W. to New Mexico was James W. Bell,
an ex-Texas Ranger who two years later was killed by the infamous Billy the
Kid during the Kid's escape from the Lincoln, New Mexico, jail. While in
Cerrillos, O. W. also met Lew Wallace, governor of the New Mexico Territory from 1878 to 1881. Wallace, who had a distinguished military career
during the Civil War and later held several important political positions in
the 1870s and 1880s, is perhaps best known as the author of the popular novel
Ben Hur: A Tale of the Christ.[39]

While in Cerrillos and Silver City, O. W. continued a romantic correspondence with Sallie Wheat, a talented singer and pianist whom he had
met in Dallas and who, coincidentally, had also attended Christian College
after O. W. left. The couple married in 1881 at the Wheat home when Sallie
was twenty-one and O. W. was twenty-eight. They returned to Silver City
on one of the first trains of the Texas and Pacific Railway to carry passengers
to the then unsettled country of southern New Mexico. There they lived for
about a year, but in 1882 O. W. lost a protracted lawsuit over his claim to the
Oasis mine, and the couple returned to Dallas.

O. W. made a living in Dallas examining land titles and estimating timber
in East Texas, but his lung condition again forced him to seek employment
in a drier climate. When in 1884 he learned of an opening in Fort Stockton
for an assistant county surveyor, he jumped at the chance, and the couple
with their young son Waldo and infant Ermine moved to the small West
Texas community, living at first in the jail, the town's only vacant building.

Clayton's mother, Sallie Wheat Williams, an accomplished pianist and concert singer in Dallas, where she met O. W. Williams.

Fort Stockton

Fort Stockton was a military garrison established in 1859 near Comanche Springs, historically a water source and camping site for Indians and later a stopping point for prospectors on their way to the California gold rush of the late 1840s.[40] Abandoned by federal troops in 1861 after Texas seceded from the Union, the garrison was reestablished in 1866–67 to protect travelers and mail coaches, cattle drives, and settlers from Indian attacks. In 1868 San Antonio entrepreneur Peter Gallagher founded the original town site of Fort Stockton as St. Gall.[41] Gallagher, an Irishman, recognized that the military garrisons in West Texas needed supplies, as would the mail stages, wagon trains, and travelers passing through the area. He and others speculated heavily in land that could be irrigated by the waters from Comanche Springs and Comanche Creek, which flowed east and north of the 160-acre tract that Gallagher platted as the town site for Saint Gall.

By 1870, the population of the Fort Stockton region of Presidio County (which included present-day Pecos County) was nearly 600 souls, mainly Irish, German, and Mexican settlers who moved to the area from San Antonio.[42] Ditches were enlarged and extended from Comanche Creek to irrigate additional land. And by 1877, settlers had formed the Pecos River Irrigation Company to use water from the Pecos River north of town to irrigate more land yet. Twelve years later, O. W. purchased an interest in the company and attempted to establish a life as a farmer and rancher.

When the Williams family arrived in Fort Stockton in 1884, the town's economy was rapidly changing from farming to ranching. Indian attacks had all but ceased and the military garrison was permanently abandoned in

Clayton with his sister Kathryn in 1895.

Fort Stockton, Texas, in 1888, four years after the Williams family arrived. The courthouse is the tall building in the background. The small square house in the lower left was once used by O. W. as an office.

1886, taking with it a source of income for the farmers and merchants in the area. The Texas and Pacific and Southern Pacific Railroads bypassed Fort Stockton by more than fifty miles, and the stagecoaches and wagon trains had all but stopped visiting the town. Soon, large cattle ranches replaced the smaller family farms along Comanche Creek, and sheep ranches soon followed. By 1890, the principal occupation of most Pecos County citizens involved either sheep or cattle.[43]

After arriving in Fort Stockton, O. W. earned much-needed income by clerking in the county courthouse. He also surveyed land in Pecos and neighboring counties for the newly arrived homesteaders and nonresident land speculators. He was elected to a two-year term as county judge in 1886, and from then until his death he would be known as "Judge Williams," "Judge," or, to granddaughter Janet Pollard, "Judgie." He served as judge again from 1892 to 1900.[44]

Sallie taught the town's first Sunday school classes in 1886 in the courthouse, which by then was also being used as a community center for dances. At Christmas, a tree was brought into the courthouse and gifts were distributed to children by Santa Claus, who was portrayed during Clayton's childhood by Herman H. Butz, co-owner of Fort Stockton's first general store. During one Christmas celebration, Clayton recalled, Santa got too close to the lighted candles on the tree, setting fire to his long white beard, which was extinguished by a well-placed bucket of water.[45]

O. W. and Sallie read the Bible at home and taught their children the Christian faith. Charter members of the Christian Church of Fort Stockton, the couple contributed to the building of several other churches in the area. O. W. was also a stalwart prohibitionist. "The Judge believed in the church and supported it substantially," said Clayton, "but was not ardent in his attendance. Mother loved her church and loved the church members. She never missed a service, and if possible she played the piano or organ and led the singing."[46]

One of Clayton's earliest memories was of riding in a covered wagon with his family in 1898 to accompany O. W. on a surveying trip to Howard's Well in Crockett County. A cat was taken along on the trip to combat any snakes that might crawl into the wagon. When three-year-old Clayton was scratched on his hands by the cat's claws, he told his mother that the "kitty has thorns in its feet."[47] On another wagon trip with his father to Fort Davis, Clayton saw his first lobos or grey wolves while the two were camping west of Fort Stockton. "It rained that evening and the lobos had killed a calf or something," Clayton recalled. "The cows were bellowing like they do when they smell blood, and a little later we heard the lobos. Father called my attention to it and the next morning we saw where the lobos had made a big track down the road right after the rain. The tracks are very similar to dog tracks but bigger. That's the only time I ever saw any lobos. The judge woke me up to hear the lobos howl. They had a low tone and looked like coyotes."[48]

Early Years on the Ranch

Five years after moving to Fort Stockton, O. W. purchased 640 acres on the south side of the Pecos River some twenty-eight miles north of town and took over the charter of the Pecos River Irrigation Company. The first irrigation project in Pecos County was built in 1869–70. It was a one and one-half mile ditch running east from Comanche Springs and irrigated three 160-acre farms. A second irrigation project was initiated in 1870 on the west side of the Pecos River northwest of Fort Stockton. There a small rock dam was placed in the river, raising the water level sufficiently to enter an irrigation ditch and irrigate 500 acres of farmland. The Pecos River Irrigation Company, which was incorporated in 1877 by Francis Rooney and other pioneer farmers, was granted a charter to dig a six-mile ditch from the river and to build another dam. The Rooneys had arrived in Pecos County in 1870 and established a 1,500-acre farm north of Fort Stockton that was irrigated by Comanche Springs. With the exception of O. W., most of the farmers living in the area served by the Pecos River Irrigation Company were from Austin and the settlement was originally called the "Austin Colony."[49]

O. W. dug more irrigation ditches in 1889 and 1890 and raised the first cotton in Pecos County. He also constructed the first road to Monahans, about twenty-five miles northwest of the ranch, on which he hauled the cotton for shipment by the Texas and Pacific Railway to the nearest gin in Colorado City, Texas. Later he helped finance the construction of a gin a few miles north of the ranch, at Grandfalls in Ward County. Powered by the flow of the Pecos River over the ten-foot falls, the gin was destroyed a few years later in a flood. The Williams farm and ranch steadily lost money, and O. W. used up his savings, sold some of the land he had acquired, and borrowed several thousand dollars from the City Bank of Austin to keep the venture going.[50]

Clayton's older brother Waldo and sister Ermine lived at the ranch from its beginning, and Ermine wrote down her observations and experiences.[51] According to Ermine, the family lived in tents for the first nine months, and these were periodically moved to follow the construction of the canal. Ten teams of mules, never completely broken to harness and difficult to handle, were used in the construction work. The family lived in one tent and stored their supplies in a second. There was also an old wooden shack with a corrugated iron roof on the property, and after the twelve miles of canal were finished, O. W. and Sallie made their permanent camp near the shack. The first year on the ranch was unusually rainy, and the family was frequently forced to retreat to the shack to sleep when the rains were heavy.

A lone Mexican cook prepared meals outside over a fire pit covered with iron bars. One day the two children took turns trying to jump over the fire pit. Waldo was successful but Ermine fell short, badly burning one of her feet. "Mother took me at once to the home of an elderly couple, Mr. and Mrs. Gooden, living nearby, and Mrs. Gooden, on seeing my burned foot, plunged it into her batch of sourdough bread by the fireplace. It must have been very good medicine, for I at once hushed my cries and the burns soon healed."

Once the main irrigation ditch was built, lateral ditches were constructed, the agricultural fields laid out, and the crops planted. A builder from Fort Stockton was hired to assemble the adobe brick ranch house, which consisted of eight rooms, and seven smaller adobe houses that housed Mexican laborers. Construction materials and groceries were hauled in from the town of Pecos, about forty miles away. The ranch was named Rancho de la Palma by the Mexican laborers after a large Spanish dagger palm on the property.

When the ranch house was completed, Sallie's furniture, including her piano, was brought down from Dallas and the family moved in. "My mother was a musician of no mean ability, and she played the piano and sang for her own entertainment as well as for the pleasure of others," Ermine wrote. "I remember that on some nights when she played and sang the hounds would gather around the front door and howl until they would get too loud for her, upon which she would sally out with a rawhide whip and chase them away."

The Williams family occupied four of the eight rooms in the ranch house, the remainder being used to store cotton seed, farm implements, and other supplies. One room served as the farm "store" where groceries, dry goods, shoes, candy, and other supplies were furnished to the workers in exchange for their labor. The room was unusual in that it contained no windows, which Ermine assumed was intentional so that no one could break in and steal the supplies. The room was locked and Sallie kept the key.

"On Friday and Saturday of each week the balancing of the accounts took place, and the Mexican laborers on those days could secure beans, bacon, coffee, brown sugar from a barrel, and sometimes cheese and rice in payment for work done. A limited amount of canned goods was available, such as canned corn, peaches, and tomatoes, and also some hard candies, but these were more or less luxuries and not dispensed except of [*sic*] holidays and other special occasions."

Some common medicines for the period—calomel (a diuretic and purgative), castor oil, quinine, etc.—were kept at the ranch, but sometimes there would be no medicine available to treat a specific illness or injury. O. W. was bothered by what he believed was rheumatism, and he made the long and arduous trip to Dallas to be treated. The doctor prescribed some green medicine and O. W. had four bottles filled, knowing that he could not obtain any more without another trip to Dallas.

When he returned to the ranch, he took his first dose. "Because it tasted so bad and the taste lingered so long in his mouth, he never took another drop of it," Ermine wrote. "My mother said it must have been very strong and good medicine, for one dose cured him completely of his rheumatism." The bottles of the foul-tasting medicine were placed in the ranch medicine chest, where they remained until an outbreak of illness, the nature of which was not recorded, spread through the ranch's population of Mexican laborers. When the supply of the usual medicines ran out, O. W. decided to give the rheumatism medicine to the sick laborers, all of whom were cured, including some who worked at a neighboring ranch.

"Word got out among the hands that Father had a very powerful and wonderful medicine of which one dose would cure a man of almost any ailment; and Mexicans came from as far away as Grandfalls to ask for a dose for a sick wife or child, and my father remarked that the only thing that kept him from becoming a famous doctor was the fact that the green medicine gave out and he was compelled to return to the old remedies of castor oil, calomel, and so forth."

A Hard Life

Pecos County was sparsely populated, and for several years the Williams family had few nearby neighbors.[52] When Kathryn was born at the ranch in 1892, Sallie was attended during the birth by a lone Mexican woman, Pancha Duran. The ranch became a stopover point for settlers traveling through Texas to New Mexico to rest their teams of horses and obtain supplies. "As a result we were constantly making new friends with the children of these families and were always distressed when they moved on," recalled Ermine. O. W. held the contract for seeing that mail was delivered twice a week from the railway station at Monahans to a neighboring ranch, and this "post office" also became a center for socializing as people came in to drop off and pick up their mail.

The road between Fort Stockton and Monahans was an important lifeline for people living in Pecos County, but there was no bridge at the point where it crossed the Pecos River. The river could be forded when it was low, but when it was running high and out of its banks, it was impossible to cross. In 1892 Ward and Pecos Counties shared the expense of building a bridge across it, but floods still blocked the crossing as water spread across the approaches to the bridge. During one flood in 1894, Waldo, then eleven, and a teamster were trapped for ten days on the north bank of the river on their way back from a supply trip to Monahans.

Due to the expense of building the irrigation ditch and the ranch house, O. W. was not able initially to stock the ranch with cattle, but he did buy one or two milk cows so that the family would have fresh milk. By the early 1890s, he had purchased 100 two-year-old heifers from a neighboring rancher and built what Ermine believed was the first branding chute in Pecos County. During the branding operation O. W. fell against the side of the chute and broke two ribs.

"At the time," Ermine wrote, "he did not realize how badly he was hurt and continued on the job until all of the cattle were rebranded. When this was done he reclined against one of the bedrolls in the back of our hack [wagon], and it fell to my lot to drive the team of half broken mules and hack back to our ranch, over very rough country made up of salt grass clumbs [clumps], while Father in pain from his rib injury criticized my driving in very pointed language. . . . I was very relieved when we arrived back at the ranch."

Working with the cattle was always an adventure for the Williams children. A few years after the birth of Kathryn, O. W. acquired a bull. He

The house O. W. built on the family ranch north of Fort Stockton. This picture was taken after the family moved into town; the house was abandoned about 1900.

warned the children to stay clear of the animal. One day the family heard the bull bawling and went to the pasture to see what was causing the commotion. There they found the bull trotting toward the house, stopping from time to time to paw the ground and bellow before continuing forward:

> The tall salt grass and brush through which he came made it impossible for us to see what was behind him and what was causing him to come toward the house. As he neared the house we finally could see Susan Kathryn's little blond head as she ran along in pursuit of the bull. We hurried out to get her and found that she was carrying a small stick which she waved and flourished with such determination that the bull was convinced. . . . Kathryn was none the worse for her experience, and the old bull seemed to survive the incident all right, although the other members of the family were quite upset for a while.

Because O. W. frequently traveled to Fort Stockton to attend to his clerking duties and, after being elected county judge in 1886, to convene court, the management of the ranch was turned over to Sallie. While she attended to ranch duties, the children received their early schooling from O. W.'s youngest sister, Anna Susan, who arrived in the fall of 1885, and at another time from his other sister Jesse. "Our school room was Father's office, located in the family portion of the ranch house, and we attended class regularly, observing school rules and school hours," Ermine recalled. Susan also taught school for $75 a month in an abandoned army barracks in Fort Stockton for three years until 1887, and Jesse taught in town in another building.[53]

In addition to cattle, the ranch acquired some hogs, but the hogs proved to be prolific breeders and were soon unmanageable. "Poor cows bogged down in the quicksand of the Pecos River and the hogs got in the habit of making a meal out of them before the help could pull them out," Clayton noted in his family's entry in a Pecos County history. "Those things and other matters soon caused several carloads of hogs to be driven to the railroad and shipped to market." The hogs could be quite vicious. Clayton recalled that numerous times he had to jump on the back of the family cow to escape a charging hog.[54]

Hogs were not the only problem animals at the ranch. Coyotes, bobcats, and skunks roamed the landscape in large numbers. The family had two dogs that kept most of the wild animals away from the ranch house, but the dogs liked to wander and were often gone for several days at a time. A good watchdog was needed to stay close to the house and warn the family should a varmint slip inside. The Williams ranch soon acquired such a dog. "Fido" joined the family when his owner, traveling from Pecos to Marathon, stopped to rest at the ranch for the night and asked the family to keep the dog. Fido proved to be an acceptable watchdog and was particularly good at cornering skunks that would sometimes slip into the ranch house at night to raid the store of supplies.

Because the family needed to keep the ranch windows and doors open at night for ventilation, the skunk invasions occurred with some regularity. Once Fido cornered a skunk in the milk room, where, before it was dispatched by Waldo, it did what skunks do when cornered. "After all these years, I vividly recall the terrible odor that permeated not only the milk room but the entire house," Ermine wrote. "It was necessary for the door to the milk room to be closed, and for months the room was used only as storage for unused articles."

As there were no screens for the windows and doors, the family could not light lamps because doing so would attract mosquitoes and other insects into the house. Smoking smudge pots were lit to keep the insects away, but the smoke bothered the children, and, weather permitting, they would climb up on the flat roof of the house to sleep. "Quite often Father would sleep on the

Clayton at age five, taken during the year he attended school in Carthage.

roof with us and we would spend an enjoyable evening as Father told us the names of the various stars and constellations. Mother, however, preferred to remain in her own bed inside the house."

Coyotes, Panthers, and Confrontation

The most difficult of the wild animals to control around the ranch were the coyotes that would frequently kill the family's chickens and raid the vegetable garden for cantaloupes and watermelons. O. W. tried poisoning the melons, but this plan backfired when a Mexican laborer became violently ill after eating one. He recovered, and O. W. stopped poisoning the melons. On other occasions he tried poisoning animal meat to get rid of the coyotes, but this plan also failed. To keep their own dogs from eating the poisoned meat, the family tied them up, and bold coyotes, no longer fearing the dogs, would come into the yard in broad daylight to kill the chickens. "We kept a loaded gun on hand at the rack close to the back door," Ermine recorded, "but it seemed always to be the case that a gun was just out of reach when a varment [sic] showed up."

Bobcats and panthers were also frequent guests, and an area near the ranch was named "Panther Draw" because of numerous panther sightings there. Ermine was once threatened by two panthers as she returned to the ranch on horseback, having fetched the family's mail from the neighboring ranch "post office." Nearing home, she called for the dogs and they chased the panthers away.[55] On another occasion, Ermine said a panther carried off a young colt that was in a corral with its mother and a mule. "The panther apparently had taken the colt from the pen, leaping over the fence with it on his back, and then had dragged it some distance where he killed it, fed on part of it and covered the remains with grass. A pack of hounds was brought over from the [neighboring] Carr ranch but after trailing the panther for several days, mostly back and forth across the Pecos River, the trail was finally lost."

As more and more farmers settled in the area, the amount of open range-land was gradually reduced, sometimes causing friction between competing interests. On one occasion when O. W. was away from the ranch, a sheep-herder cut a barbed wire fence that encircled the Williams's alfalfa field and drove his sheep into the field. Waldo and Ermine witnessed the event and raced back to the ranch house to tell their mother. Ermine said that Sallie ordered the team to be harnessed to the buckboard and took down a shotgun and cartridges. Leaving Ermine to care for her younger sister, Sallie had Waldo drive her down to the field where the fence was cut. Ermine watched from the house as Sallie talked with the sheepherder. After a few minutes, the herder drove the sheep out of the field, and he and another man repaired the fence before Sallie headed back to the ranch house. "My brother," Ermine wrote,

> told me afterwards what had taken place. When they had ar-
> rived at the cut in the fence, Mother had raised the shotgun,
> pointed it at the sheep herder and told him to come over to her.
> When he did so, she told him that his sheep had been turned
> into the field deliberately, apparently to make trouble. She said
> she would not kill him, as that would be murder; however, if
> he did not get the sheep out at once and repair the fence, she
> would start shooting the sheep and would kill as many as she
> could until she ran out of cartridges. From her manner, the
> man apparently was convinced that she meant to do as she said
> she would, and so the sheep were removed promptly and the
> fence repaired.

Throughout these difficult years of hard work and isolation, Sallie kept up her spirits through correspondence with her sister Minnie, who lived in San Antonio. Janet Pollard recalls that through the years, the sisters exchanged recipes, family gossip, and the dreams they had for their children When Sallie needed a change of scene, she would pack up the children and head for San Antonio and some social life.

"The Feud"

In the late 1890s Pecos County was still a wild and untamed country where disputes were settled frequently with fists and occasionally with guns. As county judge, O. W. found himself presiding over these confrontations in the courthouse, but one dispute became both personal and professional for the elder Williams.

To this day, it is remembered in Fort Stockton as "The Feud." It concerned the still-unsolved murder of Pecos County Sheriff A. J. Royal in 1894.[56] A comprehensive retelling of the story of The Feud is contained in Clayton's *Texas' Last Frontier: Fort Stockton and the Trans-Pecos 1861–1895*.[57] However, it deserves a briefer retelling here because the circumstances surrounding the murder provide a good reflection of the frontier society in which O. W. practiced his profession and into which Clayton was born.

A. J. (Andrew Jackson) Royal moved with his wife to Fort Stockton in 1889 after several run-ins with the law in Junction, Texas, where he operated a saloon. Royal purchased land west of town that included a farm and bought other property in Fort Stockton. In O. W.'s words, Royal was "a quarrelsome man" and was charged several times with assault, disturbing the

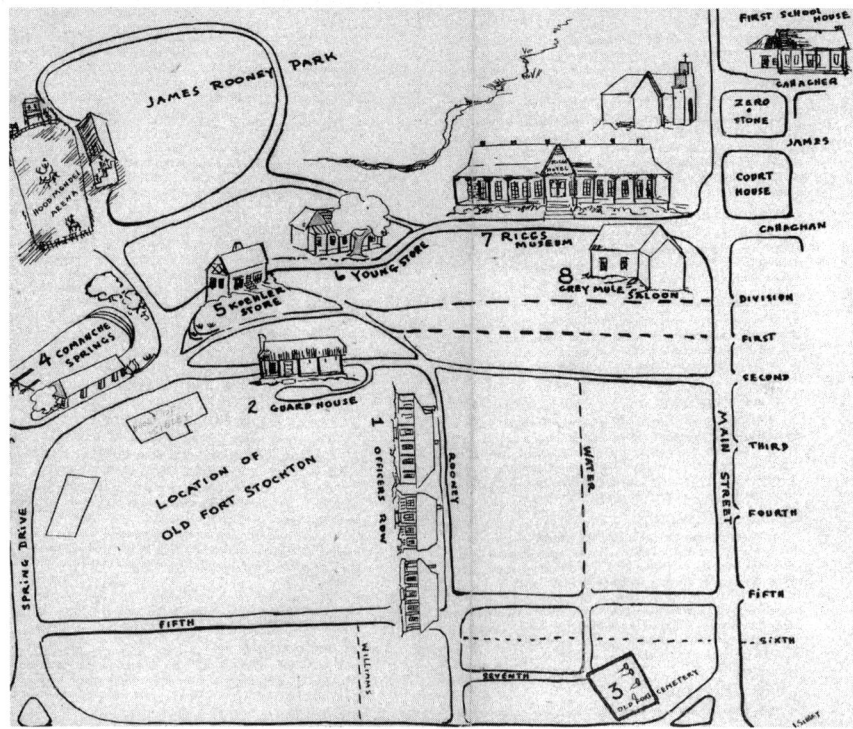

A map of historic places of interest in Fort Stockton. It shows the houses on officers' row;
Koehler's Store and the Grey Mule Saloon, around which revolved the feud involving Sheriff
A. J. Royal; and Comanche Springs. Clayton built his home in 1928 on Fifth Street near the
intersection with Rooney Street. Courtesy Fort Stockton Chamber of Commerce.

peace, and carrying a pistol, although the charges were either dismissed or never came to trial.

Royal's behavior took a decided turn for the worse in 1892 when he was charged with the May 20 murder of a Mexican who was shot in the back while standing in front of Herman Koehler's general store and saloon, a popular gathering place for the men of Fort Stockton. O. W., who had lost his judgeship in the election of 1888, served as foreman of the grand jury that investigated the charges against Royal. Royal claimed he killed the Mexican in self-defense, and in the absence of evidence to the contrary, he was acquitted. Many years later O. W. learned that several key witnesses had been threatened by Royal if they testified against him.

Royal ran for county sheriff and O. W. ran for county judge in the election of 1892. The election was contentious and Texas Rangers were called to the town to keep the peace. Despite Royal's reputation as a troublemaker, including rumors that he had stolen horses and killed other ranchers' cattle, he was elected sheriff by a vote of 193 to 137. O. W. also won his race and took office on November 17. After the election (the exact date is not known), O. W. and Royal came to blows when O. W. tried to collect money that Royal owed him for some clerical work. This fight was one of several violent incidents involving Royal and members of the Fort Stockton community. In another incident, a prisoner disappeared under suspicious circumstances after Royal escorted him out of town. A second incident resulted in a grand jury indictment against Royal in September 1893 for brutally assaulting an unarmed man by striking him three times with the butt of his pistol.

O. W. believed that this indictment created the atmosphere that ultimately led to Royal's murder. "He [Royal] became very angry at them [members of the grand jury], particularly R. L. Anderson, James Livingston, Shipton Parke and Frank Rooney . . . and Francis Rooney, a witness for [the] State," O. W. later wrote. "And from that time on he commenced a series of threats and persecutions directed against these men." O. W. later learned that when Royal gave bond for his appearance in court, he threatened to shoot the judge should the case come to trial.

However, these incidents would pale in comparison with events that began unfolding in the winter of 1894 and ultimately culminated in Royal's death that fall.

Royal and the Rooney Family

The origins of the feud between Royal and the Pecos County pioneer family of the Rooneys began in February 1894. According to Clayton, Royal and a partner started a saloon named The Grey Mule, which put them in competition with Koehler, a German immigrant who had bought his combination store and saloon in 1882.[58] Royal threatened Koehler, hit him with a walking cane, and tried to force him to close the business. Francis Joseph Rooney, the namesake and nephew of the family patriarch, heard about the threats and convinced his family and friends to boycott The Grey Mule. Learning of the boycott, Royal allegedly threatened Rooney and his friends. However, a descendent of Royal's, Sharon Bedell, told a different story. She said that Royal won Koehler's store in a card game, and that the Rooneys and others who were friends of Koehler's vowed never to go there as long as Royal owned the saloon. According to Bedell, Koehler never turned the saloon over to Royal before dying of natural causes on July 26, and the sheriff had no evidence that he had won the store.[59] Koehler was also Pecos County treasurer and at his death held more than $5,000 of county funds. O. W. appointed Francis Rooney temporary administrator of Koehler's estate. As Royal owed the estate money and was preparing a counterclaim, he was unhappy with Rooney's appointment.

On August 6 Royal was drinking with R. N. Baker, who had served on the grand jury that had indicted Royal in September 1893, and after the two exchanged words, Royal beat a very drunk Baker with his fists. During the beating Royal again threatened the other members of the grand jury as well as "those sons of Bitches" who had boycotted his saloon, and threatened to go to Koehler's store to "wipe them out." Later that day Royal went to Koehler's store alone. There he encountered James Rooney, one of Francis Rooney's four sons (the father had died in 1889). Shots were fired, and Royal later claimed that one bullet had nicked his shoulder and a second "passed through my shirt collar and knocked my pipe out of my mouth." In one version of the event, according to O. W., a witness saw Royal holding a pistol as he entered the store, and James Rooney said he heard a pistol being cocked as Royal stopped at the store's bar. In another version, Royal's granddaughter, Mary Alice Happle Townsend, wrote that her grandfather had gone to the store unarmed to buy tobacco.[60]

In her retelling of the feud, Townsend portrays her grandfather as a man who was determined to uphold the law and to rid Pecos County of crime and violence. In addition to the shooting at Koehler's saloon, Townsend described another attempt on Royal's life (as related to her by her grandmother), during which Royal was shot in the shoulder while riding his horse at night. Townsend said her grandmother pleaded with her grandfather to leave Fort Stockton, but Royal said he would leave only in a "black box." Townsend believed that her grandfather was "assassinated" for his efforts to uphold the law.[61]

Sharon Bedell wrote an account of the feud in 1974 that was later published in the *Pecos County History*. Bedell's account corresponds closely to the one written by Clayton in *Texas' Last Frontier* but is also sympathetic to Royal. She said that Royal was "a product of his time" and that, "like Andrew Jackson, Sam Houston, and other pioneers, he often had to act violently in order to maintain what he thought was right."

After leaving Kohler's store, Royal organized a few men and threatened to burn the store down and shoot the men as they came out. Francis Joseph Rooney, who was also inside, escaped, and James Rooney and W. P. Matthews, a friend of the Rooneys, surrendered. On August 7, Francis and James Rooney and Matthews each gave $3,000 bond to appear before the grand jury in September to answer Royal's charge. Meanwhile, a fast messenger on horseback reached O. W. at his ranch on August 6 with news of the events at the store. O. W. rode through the night in his buggy to Monahans, where he telegraphed W. H. Mabry, adjutant general of the Texas Rangers, to send Rangers at once to Fort Stockton. He then rode on to Francis Rooney's farm, located on the irrigated lands north of Fort Stockton, where he found the farm defended by armed men. Later that day O. W. appointed some townsmen as peace officers to keep the peace until the Rangers arrived. Five Texas Rangers reached Fort Stockton on August 9, but they left town a few days after they arrived.

Royal continued to threaten the Rooneys and their friends. Another troubling incident involved Matthews, who lost a finger from a gunshot by an unknown assailant while walking alone in town one night. "As Matthews was a crippled man, the talk was not about who fired the shot or the loss of Mathews's finger but how fast a crippled man could run," O. W. remarked. Royal also continued to work hard to remove Francis Joseph Rooney as temporary administrator of Koehler's estate. O. W. wrote two long letters to Mabry describing these events and requested that the Rangers be stationed at Fort Stockton until after the November election. They returned at the end of August.

O. W. knew that the grand jury, which convened on September 19, included several men who were either Royal's friends or indebted to him. While the grand jury indicted Royal for assault and battery against Baker and another man, it also issued a number of indictments against individuals known to disapprove of Royal. These included an indictment against O. W. for practicing law in the county without paying tax, a charge he denied, and an indictment against the Rooneys. "It was now believed that Royal had the District Attorney prejudiced in his favor," O W. wrote Mabry, "and it was clearly apparent the he could influence Grand Juries, and could intimidate honest witnesses and influence corrupt ones. We had now no hope of securing the protection of law against his attempts at our lives. Apparently there was only one of three things for any threatened man to do—via to leave the country, to kill Royal or be killed. Hard lives!"

Victor Ochoa

These events were overshadowed in October when the town of Fort Stockton and Royal became entangled with the Mexican revolutionary Victor Ochoa. Ochoa, born in Ojinaga between 1860 and 1870, led 500 rebels in an attempt to overthrow the regime of Mexican Pres. Porfirio Díaz in 1893–94. The United States, at the urging of the Mexican government, issued a warrant for Ochoa's arrest, charging that he had violated the neutrality laws. Ochoa's small army was defeated by Mexican troops in January 1894, and he escaped to the United States, where he made his way after several months to Fort Stockton. Ochoa was captured by the Texas Rangers on October 11 while he was staying at the home of a local stone mason.

According to O. W., Ochoa was popular with many Mexicans in the United States, and Royal convinced him to make a speech in the courthouse on October 21 in support of the sheriff's reelection in November. O. W. also claimed to have heard that Royal promised the fifteen or twenty Mexicans who attended the speech that if they voted for the sheriff in the November election, Royal would work to have Ochoa freed. The *El Paso Times* reported that Ochoa actually spoke in Spanish not about Royal, but about the recent Mexican revolution. Royal, who did not speak Spanish, interpreted the crowd's enthusiastic applause as support for his candidacy.

On the day of Ochoa's speech, Deputy U.S. Marshall George Scarborough arrived in Fort Stockton to take the prisoner to El Paso. According to O. W., Royal was not anxious to anger his potential voters and at first refused to deliver the prisoner, then said he would turn Ochoa over to Scarborough the next morning. Sometime during the night, about twenty men stormed the jail, subdued a substitute jailor (the principal jailer had gone home drunk), and freed Ochoa. He was captured by Rangers near Toyah three days later and tried in San Antonio. He was sentenced to two and one-half years in Kings County Penitentiary in Brooklyn, New York, for leading an unauthorized military expedition against the Republic of Mexico.[62]

Ochoa was released from prison in 1897. He lived for several years in New Jersey before becoming involved in a gold mining scheme in the state of Sinaloa, Mexico, where he was double-crossed by two men and left in the desert to die. Somehow he survived. In 1936, while walking down a street in El Paso with the town's sheriff, he spotted the two men, who drew their guns. Ochoa, who was unarmed, grabbed the sheriff's gun and shot the two men dead. At Ochoa's trial the judge suggested he return to Mexico. He is believed to have died there in 1945.[63]

November Sixth Election

Royal and his deputy, Barney Riggs, were arrested on October 30 and charged with allowing Ochoa to escape, but this did not keep Royal from continuing to make threats against his perceived adversaries. According to O. W., he was confronted by Royal at the courthouse prior to the sheriff's arrest. O. W. said Royal gazed at him for several minutes with gun in hand but said and did nothing. O. W. also heard from others that Royal had said publicly that he was going to kill O. W. Williams and several others at first chance.

On November 2, Royal charged in a warrant that O. W. had stolen three burros from him two years earlier, citing the testimony of a witness who O. W. believed was bribed. The next day O. W. was served with a new warrant in which Royal charged that O. W. and three other men smuggled stolen stock in 1890. The Rangers arrested O.W., Morgan Livingston, and Shipton Parke. The fourth man, Jim Livingston, had left town and could not be found.

The Rangers, concerned that Royal, Riggs, O. W., and the others would not receive a fair hearing in Fort Stockton, escorted them to Del Rio, about 200 miles to the southeast, for a hearing before a justice of the peace who released the men on their own recognizance. After the hearing, the Rangers advised O. W. and his colleagues to arm themselves, because they expected trouble from Royal on the train ride to Sanderson where the county election would be held the next day, November 6. There was no trouble, however, and O. W. was easily reelected county judge, while R. B. Neighbors defeated Royal for sheriff by a vote of 203 to 121.

O. W. told Clayton about one troubling incident that occurred during those anxious days. While O. W. never drank liquor, he frequently played dominoes, checkers, or chess at Koehler's store. One night someone warned him to take a different route home from the store. He did so and got home safely. He learned the next morning that another man who walked the route that O. W. normally took was confronted by someone holding a gun. The man recognized his assailant as Royal. When Royal identified the person he held at gunpoint, he said, "I'm sorry, I've got the wrong man." O. W. was so concerned about Royal's threats that he sent his wife and children to live at the Rooney farm after the election.

Royal Comes to Court

When county court convened on November 19, Royal was a major player on the court docket: he had been indicted for three assaults and was the complaining witness in five other cases. But he failed to appear. He was finally located on the morning of November 21 at his farm by the deputy sheriff and a Ranger who persuaded him to come in to the courthouse. He arrived in time for the afternoon session. One of the Rangers noted that Royal was armed with a pistol and asked O. W. if he should be disarmed. After scratching his head, O. W. replied, "I guess not, for if we did, he probably would be the only unarmed man in my court."

In a letter written on November 23 to his friend S. A. Purinton, O. W. described the atmosphere in the courthouse as Royal gave his testimony:

> He was called and appeared and testified in the first case. He
> was subdued and uneasy—but mad all through. He had his
> pistol on in Court and his Winchester [rifle] in his office—
> on conclusion of first case he went back into his office. Soon
> he was called in another case and came in and testified. Then
> again he went back to his office. This occurred four times. The
> cases being tried on persecutions—not prosecutions brought
> by him against men he did not like. So it happened that he had

congregated in the courthouse twelve or fifteen men against
most of whom it was known he had uttered recently threats of
death and all of whom he had abused or outraged in some way.
And they came there armed or at least most had arms where
they knew they could soon get them.[64]

Court was adjourned at 4 P.M. and O. W. went into his clerk's office.
"I noticed the courthouse was very quiet," he wrote Purinton, "but did not
consider it ominous until I thought of it afterward." O. W. said he heard a
voice call out "Royal" and a few seconds later heard the sound of a gunshot.
Thinking that Royal had shot someone, O. W. ran out into the courthouse
hall with his own pistol drawn, keeping an eye out for the former sheriff.
Looking toward the sheriff's office, O. W. saw a cloud of smoke in the door-
way, from which emerged Charles A. Crosby, who with Royal's support had
unsuccessfully run for county clerk in the recent election. Crosby, calling
for Sheriff Neighbors, said that Royal was dead. O. W. found Royal sitting
in a chair, his head bowed over on the desk and blood streaming down one
arm hanging loosely at his side. An examination revealed that buckshot had
entered Royal's back below his left shoulder blade and ranged up toward his
neck. His pistol was still on his body but had not been fired.[65]

Who Killed A. J. Royal?

Crosby and H. L. Hatchette, Royal's lawyer, were in Royal's office at the
time of the shooting and said they saw only some dark clothes, the barrel of what looked like a Winchester shotgun, and a sheet of flame. Neither
could identify the shooter or his voice. "There are only two Winchester
shotguns in town—one of Royal's and one of the Sergeants of the Rangers,"
O. W. wrote. "The Sergeant says his gun was locked up in his room and he
himself was at the saloon when the shot was fired."

Although Crosby was sitting at the same desk as Royal when the shooting occurred, Clayton explained Crosby's failure to identify the shooter as
the result of his being "an old man, partially deaf, and almost blind." Royal's
granddaughter, Mary Alice Happle Townsend, had a more sinister theory
about Crosby's lack of vision. She said her grandmother once said that
Crosby had been told by an unnamed party that "he would not live until
sundown if he told what he had seen."[66] Hatchette told O. W. that he was in
a corner of the room and saw only the end of the shotgun.

Who killed A. J. Royal? O. W. was convinced that someone in the courthouse "under threat of death" from Royal had killed him. O. W. believed that
Royal had successfully manipulated two grand juries and the district attorney, intimidated witnesses against him, and gotten away with cold-blooded

murder, among other crimes. "This situation forced every man under threat from Royal to realize that under existing circumstances the machinery of the law would work no protection to his life, but on the contrary was actually used as an encouragement to Royal to take life. So, I say, naturally I thought that some man under threat of death from Royal and who was present then in the Court House had done the shooting. But I saw nothing in the acts or position or situation of any one man or two men to cause me to suspect him or them more than others."

Throughout the years, rumors surfaced in Fort Stockton about several people who had an interest in killing Royal, including O. W., who firmly denied that he had anything to do with the murder. One persistent rumor was that five or six men threatened by Royal drew straws to see who would pull the trigger. What is known is that all of the individuals who had the means, motive, and opportunity to kill the sheriff went to their graves without one deathbed confession.

Royal was buried in Fort Stockton's Old Fort Cemetery on November 22. "We left the burial severely alone," O. W. wrote Purinton. The judge never bothered to send the estate a bill for the clerical work over which he and Royal had fought. When asked why, O. W. replied, "If I couldn't collect it from him, I'd surely not try to collect it from the widow."[67] The antagonism between the Williams and Royal families apparently ended with Royal's death. According to Clayton, two of Royal's daughters took music lessons from Sallie Wheat Williams.[68]

The Old Fort Cemetery in which Royal was buried is a few blocks north of O. W. Williams's house on officers' row. A marble tombstone marks his grave with the following inscription: "In Memory of A. J. Royal Born Nov. 25, 1855 Assassinated Nov. 21, 1894 Sleep husband dear and take thy rest God called thee home. He thought best. It was hard indeed to part with thee But Christ's strong arms support me. Gone but not forgotten."[69]

CHAPTER 2

Growing Up in Fort Stockton

By 1899 O. W. had established his reputation as a firm but fair judge, but he was having a difficult time making ends meet at Rancho de la Palma. Clayton recalled watching a team of horses pull his father's bedraggled cows out of the river-bottom quicksand; many did not survive. Except in times when the river flooded and destroyed O. W.'s crops, the water supply for irrigating the farm's cotton and alfalfa dried up as farms were established upstream in Texas and New Mexico. Clayton wrote that his father said it took him nearly twenty years to pay off the debts he incurred from the farm and ranch.[1]

Shortly after the family permanently left the ranch for Fort Stockton, O. W. sent Waldo, Mary Ermine, Susan Katherine, and Clayton to Carthage, Illinois, to live with his father and mother. J. C., who was born that spring, stayed in Fort Stockton. Clayton believed that the family was separated both because O. W.'s finances were at low ebb and because Carthage offered better schooling for the children.[2] Clayton stayed in Carthage for one year and then returned home; Waldo, Mary Ermine, and Kathryn remained for varying periods of time to finish school. Due to the expense, the children rarely made trips back to Fort Stockton during the first few years in Carthage, but as O. W.'s fortunes improved, the trips became more frequent.[3]

O. W.'s lack of money played a role in one of the more emotionally traumatic events of Clayton's young life, Janet Pollard recalls. "Daddy had a split-eared pet horse when he was nine years old that he regarded as 'his' horse and he rode that horse everywhere. But one day a cattle drive passed near town and some cowboys came to the house to see if they could buy any horses. My grandfather sold Clayton's horse to the cowboys and Daddy was devastated." That night young Clayton ran away and caught up with the cattle drive on foot. The cowboys refused to take him back home and told Clayton they would drop him off in the next town. Grandson Clay Pollard says his grandfather had to stay up all night and watch the herd while the

Nine-year-old Clayton riding his favorite horse. O. W. once sold the horse to cowboys driving cattle near town. A furious Clayton ran away from home, caught up with the cowboys, and stole the horse back after spending the night with the herd.

cowboys slept. "Paw Paw told us that a thunderstorm came up during the night and scattered the herd, and the boss of the cattle drive blamed my grandfather, a nine-year-old boy. My grandfather said he was so mad that he stole his horse back from the cowboys and rode it to the Downie Ranch near Sanderson, owned by family friends. They wired O. W. that Clayton was "safe and that he could stay and work cattle for a few days."[4]

By about 1902, O. W. abandoned the farm and ranch to the coyotes and panthers and concentrated on his law practice in Fort Stockton, although the family continued to own the ranch land for many years. O. W. had leased one house on officers' row, where both Clayton and J. C. were born and where he lived while in town for business or to hold court. Now moving into town permanently, he leased another house on the row. He would buy this house in 1907 for $700 and twenty-two years later Clayton would build his own home just across the street from his father.

The officers' houses were built together in a line running north and south along the west side of the fort's parade grounds. The houses had fourteen-inch-thick adobe walls that were effective at fighting the heat of summer, and each had tall windows that let in ample light, and shaded porches that provided relief from the harsh sun. The houses faced east and shared a single cottonwood tree, the only tree on the old fort grounds. When the fort was abandoned in 1881, four of the houses were torn down for lumber, but three were saved and can be seen today. The rooms were spacious, with very high ceilings. A large porch wrapped almost completely around O. W.'s house, and he added Corinthian columns that gave it a grand look. He also added a library to hold his large collection of books, which at the time of his death numbered 8,000 volumes encased in revolving bookshelves and piled on tables, floors, and any other flat, unoccupied surfaces.[5]

The porch was a perfect nursery and playground in both winter and summer. The parlor was used for Sallie's piano and music lessons, and the small, high-ceilinged room had wonderful acoustics. O. W. was also musically trained, having played the flute in college and sung in a quartet. As the children grew, they joined the ensemble. Waldo played the piano, as did Ermine and Kathryn, who also sang. Clayton had a fine singing voice and spent one summer as a teenager in Maine taking voice lessons, but he decided not to try to sing professionally when he learned how little money singers made. He also played the cornet and piano.[6] He passed his love of music along to his children and grandchildren. Janet Pollard says, "I could play the piano and Claytie was a good tenor. I sang soprano, momma sang alto, and daddy sang bass. We were a musical family and music brought all of us a lot of joy through our lives."[7] Claytie still plays the guitar and enjoys singing mariachi songs. Janet sings before groups and recorded an album of jazz standards in 2008.

The Williams's home on officers' row and the other fort buildings were located about 100 yards from Comanche Springs. A creek meandered from this large spring, and tulles (similar to the bulrushes described in the story of Moses) grew along the banks.

There were plenty of fish and turtles to be caught in the creek, and the big spring initially produced the town's drinking water, which was hauled in barrels. "The water was so clear," Janet recalls, "that you could see little minnows swimming in schools everywhere. You could open your eyes in the water and they did not burn." Comanche Springs was a beautiful oasis in the middle of the desert, and the springs and Comanche Creek formed a playground for the town's children, just as they provided a bit of respite for the early stage travelers taking the Old Spanish Trail from Saint Augustine, Florida, through San Antonio, Fort Stockton, and El Paso, and on to San Diego.

Law Practice Grows

O.W.'s law practice benefited from the growth of Pecos County cattle and sheep ranching. He located and surveyed land for homesteaders and represented many nonresident land speculators who were buying

Clayton (right) and J. C. in a homemade boat on Comanche Springs, probably around 1907. Clayton built the boat from instructions in a magazine.

up property. After the Kansas City, Mexico and Orient Railroad reached Fort Stockton in 1912, O. W. represented it in legal matters. Because of his experience in irrigation, he also became an expert in the law of surface and subsurface water rights and won lawsuits over issues involving them.

By 1915 O. W. was successful enough to construct the small building housing his law office at 100 N. Rooney Street, where he would work daily until his death in 1946. Both Clayton and J. C. later used the office, as did Claytie, who built a larger office complex nearby. The original building still stands, now marked by a plaque placed by the Texas Historical Commission in 2001.[8]

Due to his knowledge of land holdings in the area, O. W. was able to acquire much of the acreage that would later form the foundation of the Williams family fortune. According to historian S. D. Myres, O. W. registered 300 legal instruments involving land and water rights in his own name in Pecos County between 1905 and 1943, in addition to hundreds of transactions for others.[9] O. W.'s knowledge about Pecos County land was legendary. "We sometimes affectionately referred to him as a walking abstract plant,"[10] recalled one lawyer who practiced in the county. "It is almost a literal truth that he had accurately in mind both the location and

Clayton's sister Kathryn, front, riding with a neighbor near the family ranch north of Fort Stockton.

title history of nearly every tract of land in Pecos County." The same lawyer recalled asking O. W. in 1943 if he knew the address of a former Pecos County landowner named Hatch. "As I have often remarked, he had one of the most disorderly desks and one of the keenest and best organized minds that I have ever been privileged to come in contact with. . . . He walked over to a table piled about 18 inches to 2 feet deep with papers, reached deep into the pile, and pulled out and handed me a letter with a return address from Mr. Hatch dated 1913."[11]

When O. W. ran for reelection as county judge in 1900, he would sometimes take five-year-old Clayton along as he campaigned for votes. "He thought it might gain him some votes to have his son there with him,"

Clayton recalled, but he doubted the tactic garnered many votes. "Here was a little boy and a teetotaler, and most of the fellows . . . he was trying to get votes from had a hangover because his opponent had been there the day before."[12] Indeed, O. W. lost the election, which would be his last race for county judge.

Childhood Memories

Clayton's early childhood friends lived near him in the houses on officers' row and attended the small school near the town's courthouse, where they were taught by a single teacher. Children went barefoot until Christmas, when colder weather required them to put on shoes. "When we put on shoes, our feet sweated terribly and it wouldn't be long until you would have stinky feet in the winter. I remember that when one boy came in the school house everybody would know it on account of the way his feet smelled."[13]

On occasion the Williams family would buy supplies from a traveling salesman from Fort Davis. Clayton remembered that the apples sold by the salesman were particularly valued by his family, as apples were not grown in Fort Stockton. "Once, he had cloth for sale that was red, white and blue, and I asked mother to buy enough cloth to make me a suit of red, white and blue. I was real proud of that. When I dressed up, I wore it everywhere."[14]

When Clayton was eight, an early summer tornado threatened Fort Stockton, and the family watched the storm approach from their house. As the funnel cloud drew near, O. W. told his wife and children to seek shelter in the rainwater cistern. "The cistern was a pit about twelve feet deep and twelve feet across," Clayton recalled. "I happened to know it had snakes at the bottom, so I didn't follow orders. Being more afraid of snakes than storms, I avoided the cistern and sought refuge in a chicken coop, which wasn't exactly an ideal storm shelter." Fortunately, the tornado moved off to the northwest and no one was killed, but the storm picked up and killed two horses after carrying them more than a mile.[15]

Clayton remembered his Fort Stockton neighbors with fondness, recalling generous acts of kindness between families. When neighbors killed a cow, they would share the meat. Neighbors also gave him his first taste of ice cream, an expensive treat, because the ice had to be hauled to town from Pecos. "Well, every time they asked me to have an ice cream I would have it. When it was all gone, little Bob [the neighbor's son], who was two years younger than I, said, 'had more ice cream if you hadn't come.' And he was right; he would have. That was the best stuff I ever tasted."[16] Little Bob was Robert Crosby, who grew up to become a world championship roper and was inducted into the National Cowboy and Western Heritage Museum's Rodeo Hall of Fame.[17]

As young boys often do, Clayton delighted in tormenting his sister Ermine. He once hit the bottoms of Ermine's feet with a board while she was on a swing. "She was mad and mother was mad so I spent the day up in a tree hiding. That night when I did go home, mother caught me and sat on me in order to give me a spanking." When Ermine was being courted by her future husband, Charles H. Garnett, Clayton hid underneath the seat of Garnett's horse and buggy. After the pair had ridden a few miles, Garnett asked for a kiss. Clayton hollered, "Oh, give him a kiss Ermine," pinched his sister on the leg and bolted out of the buggy. The couple married in 1906 and moved to Oklahoma.[18]

Fighting

Fighting was a form of entertainment for Clayton and his boyhood pals, and they were often egged on by the older boys. "They would tell me that someone called me a liar, and they would tell that boy that I said the same thing about him. They were professionals at getting a fight going among the smaller boys. We had nothing to watch. Fighting was our entertainment." The fights could be vicious. "They'd chew on your ears, bite you anywhere they could." Once, Clayton and some of his friends were kicked out of school for fighting, but the punishment was quickly rescinded. "I told [the teacher] he might ruin our lives."[19]

O. W. expressed concern about his son's combative nature. When J. C. was a newborn, Clayton observed that his younger brother had no teeth and asked his father who knocked out the baby's teeth. O. W. speculated that if Clayton had been given the name of a specific culprit, "I think that certain boy would have had a fight on hand. For Clayton will fight, will use rocks or anything convenient. I am uneasy about him on that account."[20] Daughter Janet recalled hearing a story about Clayton and J. C. "Mother Williams went out on the front porch and she saw daddy spanking J. C. And she said, 'Clayton, stop that. That's your little brother.' And he said, 'But Momma. I would have been a much better boy if someone had spanked me.'"[21] Clayton's fighting spirit proved to be a benefit to him in later life, and his combative nature remained with him until his final days.

From a very early age, Clayton shared his parents' love of music and would sneak off to dances, often without permission. Once, while watching a dance at age nine, he was challenged by two older boys to try a drink of whiskey. When he refused, one of the boys struck him with a beer bottle, knocking him cold and requiring thirteen stitches to close the wound where the bottle had cut his nose and face. Clayton forgave the boys, who he believed meant just to scare him but not actually hurt him. "I guess they resented me being so young and going down there and seeing the dances when they were the town guys." But the experience was Clayton's first reminder that he wasn't immortal. "I prayed to the Lord as I lay there. It was one of the few times I prayed when I was young."[22]

As an adult, Clayton would reflect back on this incident and use it to teach his grandchildren a lesson about life. "My grandfather used to tell us, 'Don't run with Old Dog Trey,'" Clay Pollard recalls. "Trey was a good dog who normally didn't get into trouble and was well behaved. But one day Old Dog Trey started to run with the pack dogs that killed sheep and was shot and killed. My grandfather told us not to run with a pack that gets into trouble or you will suffer the consequences. He used as his example the time he started hanging around with some rowdy kids at a dance and he got hit over the head with the bottle. He suffered that injury because he was walking around with the kids who were rowdy—he hadn't done a thing. That was his lesson to us—do not run around with a bad crowd."[23]

When Clayton and his companions were not fighting, they learned to ride horses and mules, swam in Comanche Springs, played marbles and tops, and hunted wildcats and raccoons. "I can remember walking [home] from Comanche Springs after swimming. . . . I was barefooted and [my] feet [were] burning up on that old caliche dirt. Both the pavement and the dirt would be hot. It was lots of fun."[24] Hart Johnson recalled when Clayton, a few years older than Johnson, gave his friend a swimming lesson. Clayton first showed Johnson the motions for swimming, then threw him into the deep waters of Comanche Springs. The lesson was effective, Johnson said, because he quickly swam to the bank.[25]

Baseball

B aseball was the most popular outdoor entertainment of the time, and Clayton shared his father's love of it, frequently organizing games with friends. A marble wrapped in woolen socks became the ball, and the size of the ball depended upon the number of socks available. If a young man or his family had some extra money, a glove would be bought, but most of the boys played barehanded. A collection was taken up to buy the bats and the catcher's glove. When the town became prosperous enough to support a team of older boys and men to compete against teams from other towns, proper uniforms were designed. These were overalls cut down to bloomer size and gathered in at the waist with a band of rubber. A colored ribbon was sewn down the side of the pants, and socks the same color as the ribbon completed the uniform. Some players might be fortunate enough to purchase baseball shoes.

People came to the games from miles around, spending one day to travel by horse-drawn buggies or wagons, one day to visit and watch the game, and a third day to travel back home. The horses and wagons were driven to the sidelines where spectators sat and watched the action while holding the horses' reins or tying them to weights on the ground to prevent the horses from bolting from the excitement. During one game in 1909, sounds like gunshots erupted, and many of the horses spooked, spilling their owners from the wagons. The supposed gunshots were actually backfires marking the arrival of the first automobile in Fort Stockton. The baseball game was forgotten, and townspeople rushed to pay fifty cents each for a ride in the new contraption. The first three automobiles in town cost $1,100 apiece. They had two-cylinder engines and reached a top speed of ten to fifteen miles an hour on the rough wagon roads. According to Clayton, a trip from Fort Stockton to San Antonio, about 310 miles, would take three days, and the oil had to be changed and grease added every hundred miles or so.

In addition to conventional baseball, donkey baseball games were also organized. Players rode donkeys, dismounting to catch balls and then re-mounting to throw them. A vendor who traveled from town to town pro-vided the donkeys. "Those donkeys injured a number of fellows," Clayton recalled. "All they had to do was throw their heads down and you'd fall off in front of them because there was no mane to hold on to and nothing to keep you from falling off."[26] Clayton maintained his love of baseball throughout his life. While working for the Texon Oil and Land Company in Big Lake, Texas, in the 1920s, he took on the job of organizing the company baseball team. Many of the oil companies competed with each other and with the town baseball teams during this period, and Clayton took his job as manager so seriously that he would first try out potential employees to determine their baseball skills before offering them jobs.[27]

Two-and-a-Half

Clayton's interest in baseball was responsible for his seeing a match horse race in 1902 that he talked about for the rest of his life. The race was part of a series of events in Grandfalls, including baseball games, to celebrate the Fourth of July. Everyone living in the area was invited to attend. O. W. rode the stagecoach from Fort Stockton while seven-year-old Clayton fol-lowed on his pet horse. The race pitted a horse named Two-and-a-Half against another horse brought in from outside the area. According to Clay-ton, a cowboy had captured Two-and-a-Half while the horse was running with a group of mustangs in the late 1890s and sold him for $2.50; hence the name. The colt was raised on cow's milk and gradually grew to be a fine, beautiful bay horse, seventeen and one-half hands tall and weighing about 1,100 pounds.

Two-and-a-Half was of good temperament and made a good cow horse, unusual for a captured mustang. The horse could also outrun all the other cow ponies on the ranch where he was raised, leading some to wonder about his heritage. One story speculated that Two-and-a-Half was the son of a stallion racehorse that had been lost on the Pecos River near the present Imperial Reservoir, but his background was never determined.

When it was time for the race at Grandfalls, Two-and-a-Half was brought out pulling a surrey, which Clayton believed was an effort made by the horse's backers to make Two-and-a-Half look like a common draft horse. He would learn years later that a group of men from Midland and Grandfalls had pooled their money and bought a Kentucky race horse to run against Two-and-a-Half. A considerable amount of money was bet on the race in addition to boots, saddles, spurs, and other personal property. To the dismay of the Midland and Grandfalls contingent, Two-and-a-Half won the race. Clayton was told that the horse raced in other towns in West Texas before being sold to a man who took him to Juarez, Mexico, where Two-and-a-Half ran on a professional track for a number of years.[28]

Off to Texas A&M

By 1911 O. W. felt that Clayton, although only sixteen, was ready to apply to Texas Agricultural and Mechanical College (now Texas A&M University) near Bryan, about 100 miles northeast of Austin and nearly 500 miles east of Fort Stockton. According to Clay Pollard, his grandfather had said that O. W. had reviewed the classes Clayton would be taking in the twelfth grade at the small Fort Stockton school and found that they simply repeated the courses he had taken the previous year. O. W. decided it was time for Clayton to go to college. After spending the summer in Corpus Christi at a training school to brush up on his mathematics, Clayton passed the Texas A&M entrance exam and was admitted for a month's trial.

Texas A&M was Texas' first land-grant college and institution of higher learning. Founded in 1876 on a plain south of Bryan, the school struggled in its early years to develop facilities and academic programs, but by 1911 it was a viable college. At its heart was the Corps of Cadets, and every student who entered became a member of the Corps, subject to its rules and traditions.

Students riding the Texas and Central Railway train would pass through Bryan and be let off at "College Station," where the tracks ended and the college was located. Traveling from Fort Stockton to College Station was an arduous adventure for young Clayton, but we know little about his first trip to begin the fall term in September 1911. Waldo accompanied him in a stagecoach to Monahans, a trip of nearly a day. From there, Clayton caught a train

Clayton arrived at Texas A&M University in 1911 at age sixteen and lived in a tent with other first-year students.

to Dallas, "changed trains a time or two," and reached College Station after about three days. Because there were more students than accommodations (two new residence halls, Milner and Leggett, were under construction), he spent his first term living in a tent.

Grandsons Clay and Scott Pollard remember that their grandfather said he had to leave for A&M a week early because of O. W. "The Judge had gone to school at Harvard, and in his day the students who arrived early to campus got to take rooms closer to the university than those who arrived later," said Clay. "Judge thought the same system applied to A&M, but my grandfather said that all arriving early did was give the upperclassmen another week to haze him. And he still had to sleep in a tent, as did all of the freshmen."[29]

After the Christmas vacation, Clayton took a different route back to A&M than he had taken in the fall, because the Kansas City, Mexico and Orient Railroad was laying track to Fort Stockton and had reached the north side of the Pecos River. Waldo drove him in a wagon to the railroad construction camp, which they reached on the afternoon of December 30. In a detailed letter to his parents describing his trip back to the college for his second term, he wrote, "I expect Waldo told you of the travel to the river, of the construction train and the laying of the steel over the river. About 3:30 P.M. me and my baggage were loaded into the caboose of the train, and at 6 o'clock I left the Pecos county but my eyes and heart still lingered as long as darkness and distance would permit them."

As the construction train chugged east toward the construction camp, Clayton was overcome with homesickness. "As the sight of that country

[Pecos County] passed out of view, I sat for a long time thinking of the sad times, bad times and all kinds of times I had had in that county." But he brightened after seeing a sign that hung over a clerk's desk in the train's caboose. "Yesterday is dead—forget it. Tomorrow isn't here yet—Don't worry. Today is here—use it."

After a short trip of seven miles, the train deposited Clayton at the camp, which consisted of thirty boxcars located on a side track adjacent to the main track. He estimated that the construction crew totaled about 120 men, who were idled because the track-laying machine had recently been hit by a freight train and was out of commission. "I don't know, but it looks like the work of track laying will cease for awhile before it reaches Stockton," Clayton wrote his parents. "There is a very good chance for it to cease before it gets there because the steel is giving out. If the company has steel coming your way you may get the railroad by the last of February. If the company has no steel on the road it may be six years before you get the railroad with its service to Fort Stockton." As it turned out, the first excursion train reached Clayton's hometown in November of 1912, about eleven months after Clayton wrote his parents about the lack of construction progress.[30]

Clayton spent a rough night at the construction camp, as there was no bed and the camp food gave him indigestion. The next morning, eschewing breakfast, he boarded the construction train and traveled to Flat Rock

Clayton (right) studying with an unidentified student in Leggett Hall at Texas A&M. Like a typical men's dorm room, the walls are adorned with athletic pennants and photographs and drawings of young women.

in Rankin County, about seventy miles northeast of Fort Stockton, where he "obtained a very good dinner" and waited twenty-four hours for the next train. It was an uncomfortable wait, with only six hours of sleep in a dirty bed on New Year's Eve. For this accommodation he paid fifty cents. "The rest of the time I spent walking up and down the railroad track in order to keep warm." He found Flat Rock, named for a large rock located at the base of a hill near the tracks, a "forlorn place" consisting of a "bunch of tents and freight cars."

After twenty-four hours in Flat Rock, on New Year's Day Clayton boarded a second construction train, this one bound for Big Lake. For a payment of seventy-five cents, Clayton was allowed to ride in a boxcar "with three drunk Mexicans." Upon reaching Big Lake he boarded a train for San Angelo, where he arrived on the night of January 1. He stayed in the stately Landon Hotel, rebuilt since being destroyed by fire in 1902. There he got "one good night's sleep" before leaving San Angelo on the afternoon of January 2. He bought a ticket for Bryan for $10.65, and after four train changes he reached the college on January 3. His total expenses for the five-day trip were $18.50.[31]

Cadet Life

Clayton's academic performance at Texas A&M was not stellar. He was bothered by an eye condition that he called "granulated eyelids," which reddened his eyes and lids.[32] During his first year at A&M he was hospitalized for a month with malaria, "resulting in deficiencies in my studies.... I was not much of a student at heart." Clayton's academic performance apparently displeased O. W., as evidenced by a letter the father wrote to Clayton's younger brother J. C. in 1918. Commenting on J. C.'s poor grades at New Mexico A&M, O. W. wrote, "You seem to be traveling the same road Clayton went over—and as you seem to be headed for a failure next year—I am now disposed not to send you back."[33]

Although he had never played football, Clayton joined the team his freshman year, but he failed to pass the minimum number of hours per week and had to drop off the squad. Clay Pollard remembers his grandfather's having talked about the 1911 University of Texas-Texas A&M football game in which Texas upset the Aggies 6–0; there was a "riot" after the game. According to Texas football historian Bobby Hawthorne, the game was played in Houston's West End Park. At the end of the game a University of Texas student stole an Aggie banner hanging from the press box and had to run for his life.

After the game Aggie students roamed the streets of Houston looking for Texas fans, and the Longhorns cancelled the traditional night victory parade

Clayton in his cadet uniform with an unidentified young woman. His fellow cadets nicknamed him "Dough" because he was somewhat of a ladies man and could be seen going to or coming from the nearby town of Bryan at almost any hour.

in favor of a march in Austin the next day. According to Hawthorne, W. T. Mather, chairman of the UT Athletic Council, sent a note to his counterpart at A&M that read, "I beg to inform you that the Athletic Council of the University of Texas has decided not to enter in athletic relations with the Agricultural and Mechanical College of Texas for the year of 1912." The series was not resumed until 1915.[34]

Hazing was a serious problem at Texas A&M during Clayton's time, and like all first-year cadets, he was subjected to a number of hazing rituals concocted by the upperclassmen. Faculty efforts in 1913 to control hazing resulted in the dismissal of twenty-seven students. Members of the freshman and sophomore classes protested the dismissals, and a petition was signed by 466 students to force the administration to reinstate the dismissed cadets. A number of cadets left the campus in protest, including Clayton, who because of the walkout was dismissed from the university for six days.

"I was fool enough to walk out, but I didn't go very far," he recalled in a 1981 article published in the Texas A&M student newspaper, the *Battalion*. "I went to Bryan, and by the time my father got down [to the college]—he read about it and he didn't wait to get down here—I'd already read [a statement from the administration that said] if you agreed to abide by all of the rules of the school, you could go back to school. And I signed that up pretty soon, even before my father got there." The statement required the petitioners to pledge to abstain from hazing before the college would readmit the dismissed students.[35]

Clayton told a number of stories about his personal hazing experiences. "It is said that an idle mind is the Devil's workshop," he recalled. "At nighttime there were evidently lots of idle minds among those sophomores." One hazing ritual required Clayton to find objects that sounded plausible to the new cadet but were pure fiction. After failing to find items such as "reveille oil," "squads right blankets," and "tattoo powder," he was whipped by a sophomore with a belt, a common punishment. Another ritual required him to sing "How Dry I Am" before being drenched with a bucket of cold water. He was also required to climb a short way up a flagpole, sing "Nearer My God to Thee," and then climb a little higher and repeat the song while being nudged with a long pole.[36]

Clayton soon learned that football and school spirit ran hand-in-hand at Texas A&M. Inspired by his first pep rally, he volunteered to participate in a "pillow fight" organized by members of the football team. Two opponents were seated on a three-inch pipe that was elevated so their feet did not touch the ground. The goal was for one contestant to knock the other off with a pillow. Clayton, who was one inch under six feet tall and weighed nearly 180 pounds, won all his matches, leaving only one man to defeat. His opponent, a football player, outweighed him by forty pounds, and he hit Clayton so hard that his body spun around the pipe. Luckily, Clayton's feet struck his opponent on his way around the pipe, knocking him off as well. A rematch was held, and Clayton struck quickly, knocking the big athlete to the ground. "David had conquered Goliath," he remembered proudly.[37]

When Milner and Leggett Halls were completed about halfway through Clayton's freshman year, he moved into a ground floor room in Leggett. There it became a little easier to avoid the harassment, as Clayton quickly learned to duck out his dormitory window with a blanket whenever he heard the upperclassmen entering the building to dish out punishment. He would spend the night sleeping in the woods before returning to his room early the next morning.[38] "I mean to tell you as a fellow A&M graduate that sleeping outdoors in College Station is either going to be freezing cold or hot and muggy with mosquitoes," says grandson Clay Pollard. "That must have been some miserable hazing to make you want to go sleep outdoors in College Station."[39]

As Clayton's self-confidence grew, he demonstrated the grit and toughness that would mark his later life in the rough-and-tumble of the Texas oilfields. By the spring of his freshman year, he was fighting back against the harassment. During the corps' annual hike and encampment on the banks of the Brazos River, he roughed up a sophomore who was trying to harass him. Clayton said the sophomore threatened to return with his classmates to beat him up, but not one of them showed.[40]

Sometimes the harassment got out of hand. A favorite prank was to carefully balance a bucket of water in the transom above a cadet's dormitory room door. When the unsuspecting cadet opened the door, he would be showered with water. Janet Pollard recalls her father's telling a story about a time when the prank was pulled on Clayton. Suspecting who the culprit was, Clayton attached an electric wire to the door handle of his dorm room. After leaving the door ajar, he ran down to his fellow cadet's room, poured water on him, and returned to his now "electrified" room, slipping in without touching the knob.

"The other cadet chased Daddy back to Daddy's dorm room, touched the doorknob and got a shock," Janet recalled. "He was so mad that he went back to his room and got his saber. Daddy started running, still holding the bucket he had used to douse the other cadet. Suddenly, Daddy turned around and hit the other cadet on the head with the water bucket, knocking him out cold. How quickly a harmless prank can turn to disaster."[41]

Clay Pollard said his grandfather told another story that did end with tragic consequences. "It was the practice of the cadets on Saturday night to jump in a boxcar of a train and ride it ten miles to Bryan, where they would spend some time, probably in a bar. Then, about midnight, they would hop the train going back to College Station. My grandfather said that on one of these occasions going back to the campus, one of the young men fell under the wheels of the boxcar while jumping off and was killed. When my grandfather and his friends returned to their rooms, they discovered that this cadet was missing. When they returned to the tracks, they discovered his body.

Because the dead cadet was from Alpine and my grandfather was from Fort Stockton (the two towns are about sixty-five miles apart), he was given the sad duty of escorting the body back to the cadet's home. My grandfather told the school that he didn't know the young man very well, but the school insisted he accompany the body to Alpine. It took a week to get there by train and return as you had to go up through Dallas and then back down to Alpine. That would have been a very solemn event for a freshman in college who was only sixteen years old."[42]

By the spring of 1915, hazing incidents still occurred, but the seniors were taking more control and procedures were put in place to dismiss cadets who were guilty of egregious hazing. Clayton left Texas A&M at the end of the 1915 spring term without completing his degree in electrical engineering. According to the Texas A&M registrar's office, Clayton did not receive his degree until 1924, when he returned to school for one month to complete one course in electrical engineering and two courses in mechanical engineering. Because students at Texas A&M track each other by their graduating class, which is established four years after entrance to the university, Clayton was considered a member of the Class of 1915.

Clayton's grades for his freshman and sophomore years in most courses were Cs with a sprinkling of Bs, but they worsened during his junior and senior years. He returned to school during the 1923–1924 session and received a bachelor's degree in electrical engineering. Despite his lack of academic prowess, Clayton was an active cadet at Texas A&M and achieved the rank of cadet first lieutenant by his senior year. He was nicknamed "Dough" because he was "somewhat of a ladies' man and could be seen going to or coming from Bryan at almost any hour." Before losing his athletic eligibility due to poor grades, he played both baseball and football. His other extracurricular activities at Texas A&M included serving as athletic editor of the yearbook in 1915 and as vice-president of the West Texas Club, singing in the Glee Club, and playing Santa Claus at a Christmas reception for students and faculty in the electrical and mechanical engineering programs.[43]

Hurley, New Mexico

After leaving Texas A&M, Clayton spent a short period at the University of Texas law school at Austin in the fall of 1916, but he withdrew after a few months. "Uninterested in what seemed to be a dry subject, bothered with granulated eyelids and their treatment, and with my uncle's automobile at my disposal, my study of law suffered," he later recalled. The uncle who

The family of O. W. Williams, taken in 1915 when Clayton was twenty. Seated from left are Waldo, O. W., and Sallie Wheat Williams. Standing behind are Mary Ermine, Jesse Caleb (J. C.), Susan Kathryn, and Clayton.

had provided the automobile was William David Williams, one of O. W.'s five surviving siblings, who practiced law in Austin (and was briefly mayor of Fort Worth before joining the Texas Railroad Commission as a commissioner in 1909, a post he held until his death in 1916).[44] Clayton returned to Fort Stockton for a visit before traveling to Hurley, a small mill town in southwest New Mexico that served the Santa Rita copper mine, then owned by the Chino Copper Company. He was hired on December 2 as an electrician's helper and by February 1917 was making $112.50 a month.[45]

Clayton provided very little information about his work in his letters home. Based on an undated, handwritten chronology found in his papers, he went to the mill without a job in hand and was initially refused employment by the chief electrician. However, the mill's assistant superintendent interceded on Clayton's behalf and he was hired at the pay rate of three dollars a day. His first job involved helping erect a 24,000-volt power line and a telephone line to part of the plant.

On Christmas Day a large snowstorm blew down several power lines at the plant, and Clayton "had to go tie them up." He received a twenty-five-cent raise on January 1, and his duties involved "miscellaneous work" on the power plant, electrical conduits, and telephone lines. His narrative skips to March 1–3, when he "worked straight day and night." On March 4 he went

Cadet 1st Lt. Clayton Williams in 1915, the year he left Texas A&M before going to work at the Chino Copper mine in New Mexico.

to Silver City, New Mexico, "without leave," then on to El Paso and Juarez, Mexico, where he mysteriously "lost $100." He returned to work at Hurley on March 7 but started looking for another job, corresponding about a position with the Burro Mountain Copper Company, whose mining operation was located a few miles southwest of Silver City.[46]

Clayton regarded his work at Hurley as only a temporary stop in his career path. The war was raging in Europe and the copper business was good, providing jobs at Chino for a number of young men who worked hard and filled the company bunkhouse with dreams of grandiose get-rich schemes.

Clayton also gave himself over to such dreams when one of his bunkmates talked about a real or imagined interest in a gold mine in Panama. The mine "may get financial backing," Clayton wrote home, "and if such be the case my chances down there would be good."[47]

Clayton worked a little more than three months in Hurley, but his time there provided him with work experiences that influenced his philosophy of work throughout his life. For example, after observing one of the older employees who had many years of experience but no formal education, Clayton expressed his belief that while on-the-job learning had merit, it was not as beneficial as technical training. "Practical experience, although very good, is harped on more for the protection of the old sea dogs against the young technical men than anything else," he wrote home.[48] Clayton never lost his love of learning new technology, and years later his technical skills would pay off big in the Texas oilfields.

He also learned a thing or two about dealing with the rough life in a mining community—a life similar to the one he would live as an oilman. In his unfinished autobiography, he told the story of an encounter with a large, drunken Irishman who one night collided with Clayton's trunk, which had been placed in the hallway due to lack of storage space. After running into the trunk in the dark, the Irishman swore so loudly that Clayton strode out of his room and confronted him, with the trunk between them. The noise also prompted a large Swede to enter the fray. "The Irishman made a nasty 'crack' at the Swede and the Swede slapped him with the flat of his calloused hand, causing the flood [blood] to spurt from the Irishman's face. That one good smack sobered the Irishman up; he went to his room and a little later was arrested for disturbing the peace. The moral to that episode is: Never fight if you can keep something between you and your opponent and push on it until someone comes to your rescue. Some say 'it is better to run away'; others, 'it's better to get started away early, so you do not have to run.'"[49]

Throughout his life, Clayton would have a number of physical encounters with men who thought they saw weakness in his educated mind, courtly manner, and general good humor. He always appealed first to reason, but failing that, was not afraid to fight. And he never ran.

The Call to War

Clayton volunteered for military service after the United States declared war on Germany in April 1917, having resigned his job at Hurley in mid-March to enlist in the U.S. Army Reserves in El Paso. He does not provide any clues in his papers as to why he decided to volunteer. However, prior to America's entry into the war, many Texans were following the events in Europe with great interest. When the *Lusitania* was sunk in May

1915, a resolution was introduced in the Texas Senate to sever diplomatic relations between the United States and Germany.[50] Closer to home, Texans knew of Germany's interest in stirring up trouble between the United States and Mexico, and Texans' nationalistic pride and antagonism toward Germany were heightened after the infamous Zimmerman telegram was made public.

Arthur Zimmermann, German foreign secretary, had sent a coded telegram to the German ambassador in Washington in January 1917 proposing an alliance between Germany and Mexico. The American military's failure to capture the Mexican guerrilla leader Pancho Villa in 1916 had contributed to the German belief that the U.S. Army was not a force to be taken seriously. The Germans also overestimated Mexico's willingness to go to war with the United States. Zimmerman promised the return of Texas, Arizona, and New Mexico to Mexico if Mexico entered the war on Germany's side, encouraged Japan to join the Central Powers, and attacked the United States once America declared war against Germany. The telegram was intercepted by British intelligence and publicly released in February, resulting in outrage throughout Texas and the United States. These threats against the United States, and the sinking of three American ships by the Germans in March 1917, pushed Pres. Woodrow Wilson to declare war against Germany on April 2, 1917. Congress officially declared war four days later.[51]

Clayton's desire to volunteer may also have been stimulated by the raid that Villa made across the border into New Mexico on March 9, 1916, when he attacked the town of Columbus. Hurley and Columbus are only seventy miles apart, and the stories of this raid and two smaller raids along the Texas-Mexico border must have still been circulating in the mining town when the Zimmerman telegram was made public in 1917. Finally, as a white, college-educated male, particularly one having attended a college with a strong military tradition, Clayton was a prime candidate for the Army's first choice for its officer corps. By volunteering before being drafted, he was ensured of avoiding the grunt work of a conscripted infantry soldier.

So Clayton headed for Camp Funston Military Reservation outside of San Antonio, where in the spring of 1917 he joined thousands of other young Texas men to begin his training for a journey that would ultimately lead to the muddy fields of Flanders and the glory that was Paris.

CHAPTER 3

The Suicide Club

On a fall morning in 1917, twenty-two-year-old Clayton woke in his hotel room to the long, low moan of a ship's whistle signaling that another convoy carrying American troops and matériel to England was about to leave the peaceful harbor of Halifax, Nova Scotia. The *Kroonland* was loaded with men and supplies ultimately bound for the trenches and battlefields of Belgium and France, but it was leaving without Clayton and one of his newly commissioned companion officers.[1] As the ship slowly moved down the harbor on September 21, the young man from Fort Stockton and his comrade threw on their uniforms and headed for the docks, horrified that the *Kroonland* was sailing without them and that they would be "absent without leave."

Clayton had arrived in Halifax six days earlier. He had sailed to Halifax from Hoboken, New Jersey, after spending several days across the Hudson River in New York. There, he had waited for orders to sail to Europe after having completed artillery training at Fortress Monroe, Virginia. A newly commissioned second lieutenant in the Coastal Artillery Reserve, he was accompanied in New York by his father, and the pair spent part of their time together buying the clothes and other equipment that Clayton would need in Europe. "I had volunteered to go overseas immediately, and my new status as an officer required that I be outfitted as such with new clothes, spyglass, bedding, mosquito net, and various other things essential to a soldier during a war," recalled Clayton.[2] The military was slow to pay its new officers, and O. W. financed the new clothes and equipment. The pair took a hotel room at the Times Square Hotel for several days of shopping, seeing Broadway shows, and sightseeing before the senior Williams returned to Fort Stockton.

Clayton was ordered to Hoboken to set sail for Halifax. He and his cousin Paul Dunkel loaded a taxi on the morning of September 12 with locker trunks, bedding rolls, and valises, and headed to the Hoboken pier with 500 other young officers to begin their great adventure. Staterooms on the *Kroonland* were assigned alphabetically. Clayton found himself in a room that was several stories below deck. After securing his gear, he returned to the top deck and was assigned a deck chair for one dollar. But there was a war on, and as the ship prepared for a 4 P.M. departure, the men were ordered to return below decks so that no uniforms would be visible.

"Then with a deep long deep blast," Clayton wrote his father, "the ship started slowly out, and we felt we were leaving [the] good old U.S. for a possibly long, and certainly hazardous, experience in foreign lands. Looking back at the wharf I saw that it was a small flock indeed that watched our departure. If I sized matters correctly there were only two fathers, one mother and one sister in the lot. Not till I saw that sister weeping did I thoroughly realize what the departure meant to the people at home. And when I looked over the five hundred with me on board I could understand the reason for lamentation in five hundred homes. A finer set of young men could not be picked, and I was only too glad that my lot was cast among them."[3]

As the *Kroonland* silently glided by the Statue of Liberty, Clayton experienced the feelings of soldiers young and old who were leaving home and country to go to war. "I and a number of other boys wondered if we would ever see [the statue] or the United States, our homes and loved ones again as we moved into the unknown before us."[4]

Aboard the *Kroonland*

The trip to Halifax was Clayton's first ocean voyage, and he devoted considerable time and detail to describing the crossing for the folks back home. Heeding his father's advice to observe and record, Clayton gave his parents a descriptive tour of the *Kroonland,* pointing out various features of the ship that would be his home for the next few weeks. One popular shipboard gathering place was the "salon" at the aft end of the deck. "In it we find tables loaded with glasses of drinks, packs of cards, tobacco stubs, etc.," he wrote. "Stud poker is here the favorite game and beer the favorite drink. Tobacco [is] plentiful, very inferior in taste, but excellent in price." Aware of his parents' aversion to alcohol, Clayton ended his description of the "salon" by assuring them that "this room does not hold us long, for the dense smoke and the scent of stale beer drives any one of ordinary sensibilities out for fresh air."[5]

Shipboard gambling was also an occupation in the second-class dining room. Clayton noted that "every gambling game in the world, or at least all that are known to these boys, has been played in this room" and wrote that one fellow officer had won $400 on one throw of dice. Again, so as not to alarm his conservative parents, Clayton added that while it was amusing to watch the gambling games, "my experience in them is small, my skill is still smaller, and my pocket-book is smallest, so I keep out. I am satisfied with the thrill that comes to a 'Looker on in Vienna.'"[6]

The *Kroonland* soon anchored in Halifax harbor and awaited the arrival of other ships and the destroyer that would escort the convoy to England. The local community was having a stock show, and the officers were permitted to go ashore in lifeboats to take in the sights. Clayton recalled that after two or three days at anchor, he and a companion "made the trip to view one of the loveliest fairs I had ever seen." Relieved at being away from military life for a few hours, the pair missed getting on the last lifeboat that day for the return to the ship. Knowing that the lifeboats would return the next day for another load of officers, the pair rented a hotel room on the night of September 20, waking the next morning to the *Kroonland*'s departing whistle.

"We hurriedly paid the hotel bills [and] rushed down to the wharf where small boats were arranging to go out in the bay. The first one we asked to take us out to the *Kroonland* turned us down but the second one was to order for us because it was carrying a lately ordered number of Sam-Brown [*sic*] belts to our very ship." A Sam Browne was a belt worn by officers in the American Expeditionary Forces (AEF) and the French and British armies. Named for the British officer who designed it while fighting in India during the 1850s, it was a wide belt that circled the waist with a narrower strap that passed diagonally over the right shoulder.[7] "We got to the ship as it was slowly moving along, climbed up its rope-ladder with some Sam-Brown belts over our shoulders and acted as if that were exactly what we were supposed to be doing. Fortunately for us, there were no repercussions."[8]

Clayton's journey to the deck of the *Kroonland* had so far been a great adventure for the young man. He had earned his commission one month earlier as one of the "ninety-day wonders," volunteers given three months of training before being commissioned as junior officers to lead and train newly formed American divisions headed to war in Europe. These junior officers were needed because the American military in 1917 was ill prepared to fight a global conflict. All volunteer since the Civil War, the army was primarily responsible for protecting settlers from Indian incursions and required few troops. There was no mechanism for mass recruitment, and volunteer recruitment was expensive. Recent engagements with foreign powers in the Spanish American War of 1898 and against the Mexican guerrilla leader Pancho Villa in 1916 had demonstrated that the army was severely undermanned, undertrained, and poorly equipped.[9]

The First Campers

W hen President Wilson declared war in April 1917, he realized that it was unrealistic for America to enter the new conflict with a volunteer army, having observed how the British had turned to conscription in 1916 to absorb the tremendous loss of manpower in the early war years. Congress passed the Selective Service Act of 1917 at Wilson's request, and by June the army was calling for 687,000 men to be added to the existing force of about 220,000. The army projected that it would ultimately need about 200,000 officers to train the new troops, about half of whom would be recruited or drafted from scratch, and it built sixteen hastily assembled officer training camps at thirteen military posts around the nation.[10]

Before Clayton could begin his military training, he had to pass a physical examination in San Antonio, and the officer who examined him initially hesitated because of Clayton's eye condition.[11] After two visits to an eye doctor, he was cleared for military service and reported to Camp Funston Military Reservation, about sixteen miles northwest of the city. Here, in Leon Springs, the military had established the First Officers Training Camp. In May 1917, in the dusty ranchland of central Texas, Clayton joined more than 3,000 other young Texans who had volunteered for officer training. Applicants were required to have at least a high school education, "to be as physically perfect as mortals can reasonably be, and to have their character and good habits vouched for by three leading citizens in their community." The army had originally planned to form fifteen companies of 165 men each at the camp, enough officers for an infantry division plus attrition, but the volunteer response was so great that eighteen companies were organized.[12]

"I was placed in Company Seven," Clayton recalled:

> My lieutenant was named Levine, and he told a funny story in the beginning to his company. He said he had just been on the Rio Grande with a bunch of green militia men that had come from all over the United States to get a little experience during the Pancho Villa trouble.
>
> One night the colonel thought that he better check up on some of his sentinels. He had them on railroad bridges and other various places of strategic importance. One of the sentinels said 'halt' and the colonel halted. Then the sentinel said 'advance and give your counter sign.' The colonel advanced again and gave the counter sign. As he started forward, the sentinel said 'halt' again and the colonel halted again. Then he had to give the counter sign again and the colonel thought there

Clayton (first row, fourth from left) with his fellow "First Campers" at the First Officers Training Camp at Camp Funston Military Reservation near San Antonio. Clayton volunteered for the war in the spring of 1917. Throughout his life, he attended reunions of the First Campers. Courtesy Texas Military Forces Museum, Austin.

was something wrong here. He said 'sentinel, what are your special instructions?' The sentinel answered 'halt them two times and if they just keep coming, shoot hell out of them.'[13]

The "First Campers" at Camp Funston, who came from all walks of Texas life—farmers and ranchers, bankers, lawyers, teachers—were overwhelmed by three months of intense military training. The first blush of patriotism that had inspired them to enlist was soon replaced by the realities of military life. During the first month, their every waking minute was consumed with learning the skills that would enable them to lead men into combat. From early in the morning until late at night, they absorbed instructions during close-order drill on how to perform and give commands; how to aim, sight, and fieldstrip a Springfield rifle; how to administer first aid; and how to select bivouac sites, sterilize water, lay out a kitchen area, and design an incinerator or garbage pit. Many were mustered out for physical reasons, ineptitude, inability to adjust to the new environment, or for situations at home. By June 18, there were fewer than 2,000 men left at Camp Funston.

Coast Artillery School

At the end of the first phase of instruction, men selected for technical training as engineers left Camp Funston for Fort Leavenworth, Kansas, and others were assigned to the Aviation Ground School at the University of Texas at Austin. Clayton volunteered for the Coast Artillery and was sent with other volunteers to Fortress Monroe, Virginia. Those who remained at Camp Funston were organized into nine infantry companies, three field artillery batteries, and one troop of cavalry.[14] The Coast Artillery was primarily responsible for operating fixed-gun batteries in coastal fortifications as a defense against naval attacks. "I volunteered to go up there (Fortress Monroe) because I thought I would do a little better with the big guns because of my electrical engineering training," Clayton recalled. "So they bundled us all up who wanted to go up there. Included in that bunch was my cousin Ned Holland and they chartered us a whole railroad car."[15]

Fortress Monroe is located near Hampton, Virginia, at Old Point Comfort on the tip of the Virginia Peninsula where the James River meets the Chesapeake Bay. It was constructed between 1819 and 1834 and named in honor of Pres. James Monroe. Completely surrounded by a moat, the six-sided stone fort in 1824 became the army's first service school for instruction in artillery. Instruction was held intermittently until the Spanish-American War necessitated a suspension of all school duty. The War Department reopened the school in 1900, and the separation of Coast and Field Artilleries in 1907 brought about a reorganization of the school as the Coast Artillery School. Instructional courses included artillery and gun defense; electricity and mine defense, principally submarine defense; and specialized training for electricians, firemen, and master gunners. The school was operated at Fortress Monroe until 1946, when it was moved to California. The fort is on the National Register of Historic Places.[16]

Clayton was among the 1,200 officer candidates who entered the Coast Artillery School in June 1917. The demand for artillery officers placed a great burden on the school, which normally trained from forty to fifty officers in a year. The school's wartime objectives were to provide two months of intensive training so that these officers could command heavy artillery in the field and in harbor defense. When training was complete, 200 of the candidates were commissioned in the Coast Artillery Corps and 566, including Clayton, in the Coast Artillery Reserve Corps.[17] Clayton was visited by his father during the latter part of his training at Fortress Monroe and, as Clayton related later, it was a good thing O. W. came to Virginia. "Somebody had put on the bulletin board that I and two other boys were not going to get a commission. My father went right to the Colonel and found out that was not true. Meanwhile, those other two boys had skipped out to get them another job. No one ever looked for the perpetrator of that dirty deed."[18]

Sailing to Europe

With Clayton onboard, the *Kroonland* made the crossing to England without incident, accompanied by thirteen troop ships and a destroyer. Some ships were carrying soldiers from Australia and New Zealand. Less than three months after Clayton's ship sailed out of Halifax harbor to

join its convoy, the harbor was destroyed by what was then the most powerful explosion ever seen in the world. On the morning of December 6, 1917,
the French ship *Mont Blanc,* packed with 2,925 tons of explosives being
shipped to Europe, collided with the Belgian relief ship *Imo,* caught fire, and
exploded, killing more than 2,000 and wounding another 6,000.[19]

Clayton's Atlantic crossing was not without danger, as the threat from
German submarines was readily apparent. A naval lieutenant on board told
the men that on a previous crossing he had seen a number of dead horses
floating in the water, killed when a transport was torpedoed.[20] The convoy
followed strict procedures for avoiding submarines, including closing and
masking the portholes at night and prohibiting all smoking on deck. Clayton recalled that the ships communicated changes of speed and formation by
flashing signals to each other. "The leading ship is the guide, and by means
of a small light aft, which is confined to a line within the scope of sight of
its followers, the ships following keep true to the leader's course. So here are
20,000 men in 14 large ships, loaded also with several tens of millions worth
of cargo, following in the dark a shaft of light no bigger than an ordinary arc
lamp, aimed at them from the leader."[21]

About the third day of the crossing, Clayton woke to rough weather
and a rougher stomach, and joined his fellow seasick comrades above deck.
"Although I still felt badly I believe that the hard-hearted enjoyment I got
during the next half hour in watching these performances did a great deal
to restore me to a normal feeling. The poor devils would sit around with an
expression of awful woe, relieved now and then by a furious dash for the
rail.... One man was asked if he had a weak stomach. 'No,' he said. 'I am
throwing it as far as the rest of 'em.' Another unfortunate fellow made the
statement, 'If my folks shall ever want to see me again, they must come over,
for I'll never cross the d—d [ocean] again.'"[22]

As the convoy entered the war zone around England, the danger from
submarine attack heightened. The men were told that sentries had been
ordered to shoot anyone showing a light on deck after dark. The officers
on guard were instructed not to be too quick to raise a general alarm about
suspicious objects in the water because floating wreckage was plentiful in
the war zone. "Under such warnings and instruction we rode our boats with
uneasiness as they raced through the war zone," Clayton wrote home.[23]

It is apparent from Clayton's letters home and his later writing that he
and his comrades were very concerned about the submarine threat, but there
was always an amusing incident occurring on board to lighten the gravity
of the crossing. One took place while the *Kroonland*'s machine guns were
being test-fired. An officer whose room was well below decks was afraid of
being trapped should the ship be attacked. He fled up the stairs to the top

deck, fully outfitted in a life belt and heavy overcoat. "He went out so fast that only two men in the room noticed him. But you can depend on it that we were all watching for his return, and when he came slipping back hoping to escape notice, there was an unnatural silence all over until he got through, when the crowd turned loose so heavily that the guns could no longer be heard. Through all the sense of danger felt by every one," Clayton wrote, "the humor of this thing was too much for good healthy young fellows to keep the brakes on."[24]

Clayton told another story about a doctor in their group who was extremely nervous during the crossing. "He had some of the crewmen bring him up a barrel and fill it full of water. He got in there with his lifesaving jacket on to be sure it would float him. But that same doctor worried all the way and when he got [to England] he was white headed."[25]

The men were ordered to keep their life preservers with them throughout the war-zone passage. Many slept in their wool clothing, thinking this would afford them some protection from the cold sea water should the ship be sunk.[26] Stories about past experiences with German submarines made the rounds of the ship. They were no doubt generated by the sailors, including one such "brush" that the *Kroonland* had had with a sub during an earlier crossing:

> One day a periscope was sighted several hundred yards to one side of one of our ships. In a short time it disappeared, after having probably noted the distance to the ship, its course and speed. To meet this, the ship's captain, instead of speeding up, reversed the propellers and stopped the ship, fronting the last above-sea position of the 'sub.' In a few minutes the sub, not understanding the situation, arose for observation, and emerged under the stern of the ship, and right into the ship's propellers. The ship's propellers were started at top speed at once, and after some revolutions were stopped. There were then visible at the stern some bubbles of air in the water with a great amount of oil spread on the surface. The propeller was found badly broken in places. Nothing more was seen of the sub, and no one on the ship knew absolutely that it sank, but few doubted it's [*sic*] fate.[27]

"Those Damnable Fifty-Eight-Twos"

On the afternoon of September 29, the convoy was met by American and British destroyers and escorted to various coastal ports, including Liverpool, where the *Kroonland* docked on October 2.[28] Clayton was impressed by the large numbers of children he saw on the docks diving into the harbor to retrieve pennies thrown by the Americans. On the night of October 3 he crossed England to Southampton where he boarded the *Londonderry* and crossed the channel to Le Havre, France.

"The roughest ride I have ever had was laid on that boat," wrote Clayton, who had managed to secure a spot below decks. "As I sat down in that close and stuffy but warm and dry hold, I felt sorry for those poor boys who had only standing room on the exposed upper deck. . . . But my sympathy did not lead me far enough to offer to exchange places with any one of them." While below deck, Clayton joined a group of several British officers who were returning to the front after recovering from wounds in England. "They were engaged in relating their various experiences in the war. It was exceedingly interesting to me, looking forward to the same field of action."[29]

Clayton was in the first wave of American combat troops to reach France, and by the end of 1917, about 175,000 Americans were in the country. The first American troops arrived in Europe about three months before Clayton. By October, the First Division was at the front lines in northern France, and the first American combat deaths occurred in France in November.[30] Clayton, along with fifty of his Coast Artillery comrades, reported for training at the French Trench Artillery School at Bourges in central France. Trench warfare was new for the U.S. Army, and the goal of the school was to provide American officers with enough information about trench artillery to train the American troops coming in behind them. Due to a number of manufacturing problems, the army was unable to provide much artillery. The French provided the majority of the heavy weapons and artillery, with the exception of trench mortars, most of which were provided by the British. This was apparently due to a belief among the AEF command that French trench mortars were unreliable, although Clayton did not speak to this issue in either his letters or his later writing or in discussions with his family.[31]

While waiting to take a train from Le Havre to Bourges, the young Texan once again almost missed his chance at the war. After marching five and one-half miles through Le Havre and into the surrounding countryside, Clayton and his comrades were ordered into an English rest camp composed of tents pitched in a field of heavy mud. Here the men got a meal and borrowed some warm blankets. "After some hours of rest on dry blankets we were notified at 8:00 P.M. to do up our baggage ready to start at any moment," Clayton wrote. "We did so, and waited further orders; tired at last of waiting, I laid down on my borrowed blankets and dropped off asleep." When Clayton woke at 4 A.M., his comrades were gone. "I realized that I might be called a deserter, although it seemed to me that I was the party deserted, so I hurriedly rolled up and returned my borrowed blankets, and rushed down just in time to catch the last baggage truck bound for Harve." Upon rejoining his comrades at the train station, Clayton learned they had gone down to the station on trucks the night before and had stood all night in the rain waiting for a French train.[32]

Clayton's selection for trench artillery was arbitrary. Other officers were assigned to heavy and field artillery, antiaircraft guns, and machine guns. "Well," he wrote home, "I drew my lot in the Trench Mortars, and here I be, trying to do the best I can in a service strange to me."[33] He expressed some envy of officers who were assigned to antiaircraft artillery. "It will be an accident if any [antiaircraft officer] ever gets touched. It is calculated that these guns will hit an aeroplane about once in 15,000 shots, but that is sufficient it is believed to keep the 'Bouche' planes at such an altitude that they cannot make effective observation or photographs. The aircraft guns are carried around in autos, and the officers ride in good cars."[34]

Motorized artillery was first developed by Germany in 1909 and consisted of a 75-mm twelve-pound gun on a motortruck mounting. After the war began, the French and British mounted antiaircraft guns on trucks and trailers, and by 1917 the AEF had also developed a trailered gun. Germany and the Allies developed fixed antiaircraft defenses at locations that might attract enemy aircraft, and considerable effort was spent on developing systems for aiming the guns and calculating artillery shell fuse lengths, elevations, and other elements critical to shooting down aircraft successfully.[35]

For Clayton and other Coast Artillery officers, assignment to trench mortars was due less to the luck of the draw than to who was doing the drawing. According to one officer who arrived in France with Clayton, "An improvised method was adopted by which the necessary number of officers were to be selected. A fake drawing of little papers was the scheme. One young officer volunteered to handle this, and no further mention need be made here other than to call attention to the fact that his friends, like himself, drew heavy artillery, while those whom he did not like or know were placed on the trench artillery list." Clayton also noted in his letters home that the lieutenant in charge of the drawing and his friends were all assigned to heavy artillery.[36]

Several concerns fed the reluctance of officers to volunteer for trench mortar duty. A mortar is essentially a short, stumpy tube that fires a projectile at a steep angle (higher than forty-five degrees) so that it falls straight down on the enemy. Because trench mortarmen were by necessity located at the front, officers chosen for this duty were exposed to heavy enemy fire. In addition, efforts by Allied troops to respond to Germany's use of mortars in the early years of the war were often ill conceived and dangerous to the firing

*Commissioned a first lieutenant in the Coast Artillery Reserve
Corps in 1917, Clayton was in the first wave of American troops sent
to France.*

crews. Stories about Allied mortar failures were still talked about when the
first American troops entered the war.[37]

Reinforcing these concerns was the practice among Allied troops of referring to those assigned to trench artillery as members of the "Suicide Club."[38]
Clayton told the story of one fellow officer whom he nicknamed "Job." Mis-
interpreting his Bible, Clayton said he named the officer Job because he was
a continual complainer.[39] Job had no sooner received his trench-artillery assignment than "a Red Cross British train came carrying seriously wounded

soldiers into Le Havre to place three (soldiers) on the very boat we had
come in on. Job questioned a few of the wounded as to what kind of army
actually was close by the Trench Mortar men. He got many exaggerated
answers, for those veterans knew at once he was a 'greenhorn.' One told
him, 'it was a suicide outfit and no one lasted very long on the front while
shooting those missiles, for both the Germans and the Allied soldiers hated
its members.'"[40]

Despite the danger, Clayton was proud of his assignment and assured his
parents that trench artillery was "in a safer place than the trench infantry
and the field artillery" and was "the most effective branch of the artillery
service." He plunged into his training with gusto and received instruction in
trench mortars, machine guns, illumination rockets, gas, and the digging of
emplacements and trenches. "Here I came back once more to student life,"
Clayton wrote home. He also helped write a trench mortar training pam-
phlet for use by the U.S. Army.[41]

The French were happy to see the Americans and went out of their way
to make camp life as pleasant as possible for their guests. "Our pay was slow
in getting to us," wrote Clayton, "and had not the French officers been exceedingly kind to us, we might actually have suffered for food. Upon our arrival here, they made it a special care to see that our every need was promptly
attended to [and] provided us with barracks, men and servants, even to the
extent of giving us orderlies to make up our beds, to light our fires and even
to black our shoes."[42] The French officers also gave the Americans a dinner
party shortly after their arrival, complete with menu cards featuring drawings that illustrated "the bond of friendship between the French and Uncle
Sam." As one American officer noted, the dinner was "an effort on the part
of all concerned to raise the price of champagne in Bourges, resulting in an
attempt to paint the town red, white, and blue, and ending in a snake-dance
up to the Club, where we found the guard already turned out for a strike,
riot, or what not."[43]

Some Americans drank a little too deeply of the pleasures that Bourges
had to offer. One officer, after a night in the cafés, staggered back to the large
darkened room where Clayton and his companions were asleep. "In trying to
find his way around in the dark, he walked into the large pot-bellied stove,
which heated the entire room," Clayton wrote. "Somehow, he got the idea
he must defend himself and [thought that an] attack by him might be the
best defense. We were all awakened when he started striking out in front
of him in a similar manner to which one might imagine Don Quixote, the
would-be knight of chivalry, attacked the windmill. When we got the lights
on, the lieutenant was seen covered with soot and struggling among several
stove pipes he had knocked loose from their connectors."[44]

While at the French school in Bourges, Clayton described an amusing incident involving horses and the officer he had nicknamed Job. AEF officers were notified that they were in the class of mounted officers and were expected to know how to ride. The officers chose their mounts from some 200 horses that were "old leftovers" from the French front, and Clayton noted that Job chose an animal "of uncertain temper" that would not make a good saddle horse. Clayton chose a pony, reasoning that it would be easier to mount and dismount the small animal. On the ride back to camp, Clayton was challenged by another officer to race, and Job joined in. As the three officers sped back, they approached a ravine that bordered a field freshly fertilized with human manure. Fearing that he would be badly thrown when the horse jumped the ravine, Job prepared to leap off. "He was a little slow, however," Clayton recalled. "The horse had jumped the ravine and risen to the open field before he launched himself overboard in spread-eagle fashion. And he had not made a good selection of a landing place. He fell in a choice collection of the fertilizer and covered himself not with glory, but with the unpleasant elements that go to make good French crops."[45]

The officers were trained in Bourges to use a variety of French mortars ranging from the primitive 58-mm No. 1, which was first put into use in 1915 and had a range of several hundred yards, to the very large 240-mm mortar that could fire a projectile 2,400 yards. The mortar was a smooth metal tube fixed to a base plate. A bomb with an impact-sensitive cartridge was dropped into the tube, where it would make contact with a firing pin at the base and fire the bomb out of the tube toward its target. "Our studies covered the range, dispersion of shots, charges, bombs, fuses, etc.," Clayton wrote. "As they are for service at close range to the enemy, these guns are much more certain to hit their object than the large long-distance guns and can carry a much larger explosive charge."[46]

There were many varieties and types of mortars used by the Allies during the war. The most popular was the British Stokes mortar, named for its inventor F. W. C. Stokes, who later became Sir Wilfred Stokes KBE.[47] In addition to the 58-mm No. 1, many of the first American trench artillery officers trained on the 58-mm No. 2, which was developed about the same time as the No. 1. The Americans also used the six-inch Newton mortar and the 240-mm heavy mortar. The French 58 mm No. 2 inspired a song popular among the American trench artillerymen. The author is unknown.

> You can sing of the glory of death on the lyre,
> Of ar-e-o-planes for their swank,
> Prefer to pass out to the fumes of the gas,
> Or give up your life in a tank.

But I'll sing to you of a job that is new,
A job worse than any disease,
A job where your coffin is ever in view,
Or you can die any death that you please.

Working on fifty-eight-twos,
You can die any death that you choose.
Get hit by a shell, the gas gives you hell,
Grenades wake you up when you snooze.
But your family is proud of you now,
They are sure to collect that ten thou.'
Ten days to each man is the average span,
Working on mortars, those jolly trench mortars,
Those damnable fifty-eight-twos.[48]

The training could be dangerous, and Clayton described one training mishap that had the makings of a major disaster. "On one occasion a charge of wet powder was put in a mortar by some oversight, and the bomb was loaded and timed with an instantaneous fuse. The result might have been disastrous for the bomb was thrown only about 10 feet in front of the gun. But it happened that the shock was not sufficient to set off the time fuse. We were ignorant then of the danger, and the only persons to act promptly were the French sergeant, who recognized the situation, and Lubin [a fellow student], who had a suspicion of it, and on that suspicion made the 100 yards between the mortar and shelter in 8 seconds, and he did not have a good start at that."[49] In another training mishap, Clayton almost lost a finger when a cannon barrel pinched his hand, and he termed himself, no doubt tongue-in-cheek, the "first American in the new branch of the service to be wounded in France."[50]

Despite the dangers of training and some bouts of homesickness, Clayton seemed to enjoy his time in Bourges. In letters home he described the beauty of the Gothic Cathedral of Saint Étienne, constructed between 1195 and 1255. He was particularly struck by the tone of the cathedral bell. "It must be the fabric of a master builder in bronze, for the sound that comes out of the old belfry is wonderfully soft and sweet, not like the brassy clang of our bells, and yet it carries remarkably."[51]

He also spent some time walking through the city and noted the "old, well built, solidly constructed houses" with their "strongly barricaded" doors and windows, and the cobblestoned streets that were "difficult for a prairie raised Texan to walk over." In his final letter to his parents about Bourges, he told a joke on himself about the cobblestones, relating how he heard a horse and buggy coming up behind him and jumped to the side of the street, only to discover that "my big horse and carriage with its reckless John of a driver was only a little 7 or 8 year old Street gamin, hurriedly walking up street in a pair of large and loud wooden shoes. And the laugh was on me."[52]

CHAPTER 4

French Friends and New Adventures

Clayton finished his training at the French Trench Artillery School on November 17 and received orders to help establish the AEF Trench Mortar School at Fort de la Bonnelle, near the city of Langres, southeast of Paris and just north of Dijon. The school combined the American officers who had trained at Bourges with ten American officers who had attended the British Second Army Trench Artillery School at Saint Omer, France, a total of sixty men, as well as two French officers for liaison and instruction. Capt. (later Col.) Forrest E. Williford, CAC, was named director of the new school.[1]

The road to Fort de la Bonnelle led through Paris, and Clayton and his comrades took advantage of their downtime between training schools to spend a day sampling the sights of the "City of Light." Their adjustment to Parisian customs, however, was a bit awkward, and the first sense that things were different began when Clayton and his two fellow officers arrived in Paris by train. "When the train stopped, everybody began to tumble off and run away," Clayton later wrote his father.

> I was not quite certain whether I was in a depot in Paris or in a tunnel with an alarm of "fire" on hand. "In doubt, play trumps," so I followed the lead of the crowd, and I assure you it moved quite fast.
>
> Now a peculiar feature of railway travel here is that the ticket is collected after your journey is over and you are going out of the gates. Well, in my mad rush to keep up with the crowd, I ran clear and clean over the Conductor at the gate, and you know he had to run *some* to get the ticket. When I got to the open, there was another surprise. Instead of finding 50 taxi drivers yelling at me to ride with them, as I thought I had a right to expect in any world metropolis, I saw some taxis

lined up against the curb, with drivers standing in butler-like attitudes of solemn silence. Further, I saw my late passenger comrades madly chasing up and down the "Rues" after running taxis frantically waving arms and valises after the retiring cars.

After approaching several drivers, Clayton discovered that all had been previously hired and were waiting for their passengers. "This peculiar conduct results from a shortage of gasoline," he surmised. "Taxis only come and go where they have already secured passengers or engagements, and they never hang around anywhere on the chance of passengers." Clayton was able to hire a "street urchin" to run up and down the street and engage a taxi for the three soldiers.[2]

After settling in at the Hotel Continental, Clayton and his comrades made arrangements to locate a French family that would ultimately provide the young Texan with many enjoyable diversions during his time in France. As Clayton later wrote his parents, the trio had been given the address by a French officer who said a woman wanted to meet some American soldiers and become a "godmother" to them. The French tradition of the *marraines de guerre,* or "godmothers of war," was usually extended to French soldiers who were either cut off from their families by the Germans in northern France or who had no families at all. The "godmother" wrote letters to her "godson" at the front, sent necessary materials, and provided him with food and a place to stay while he was on "permission" or leave of absence.[3]

The French family that Clayton and his comrades met that day consisted of Pierre Jean Gaubert and his wife, their adult son Georges, and their married daughter, Suzanne Sykes-Gaubert, who had requested to meet the American soldiers. "Our godmother [whom Clayton referred to as Madame Sykes] is pretty, the wife of an American and speaks excellent English," Clayton wrote home, noting that she also had three children. "Her father is a munition manufacturer (chlorine gas) who also uses good English."[4] The Gauberts and their daughter entertained the three American doughboys royally during their first visit to Paris. "After lunch with them, we were carried in an auto over Paris to different places of interest such as Notre Dame, the Eiffel Tower, Napoleon's Tomb, etc. That evening we were carried to the country home of the son to an afternoon coffee, spiked I judged with rum, and as it was misty and cold, it went very nicely to us after our ride."

On their way back to Paris, the family had to stop their motorcar to have the gasoline measured and a tax paid on it. The French government stopped issuing gasoline to private automobiles shortly thereafter, and Clayton, apparently anticipating that action, remarked that his next trip with the family

Clayton with Madame Gaubert. During his time in France, the young soldier was befriended by the family of Pierre Jean Gaubert in the tradition of the marraines de guerre *or "godmothers of war," who became second families to soldiers.*

"may not be by auto."[5] Returning to the Gaubert home in the Paris suburb of Saint-Denis, the Americans were treated to a French dinner featuring a variety of fine wines (including one 1880 vintage) and three cognacs.

"There were many peculiar French dishes with which I was not familiar, so I just waited the lead of our hosts as to them. After dinner, cigars and cigarettes [were provided] and we all went to the reception room where a piano was opened and we all had nice music and conversation." The Americans said their farewells at 11 P.M. and returned to the hotel, from which they departed at 8 the next morning for the newly created AEF Trench Mortar School at Fort de la Bonnelle. Clayton told his father that his day in Paris left him "feeling certain that France had some extremely nice and polite people."[6]

In subsequent trips to Paris throughout the war, Clayton would always take time to visit with the Gauberts. He said that on one of his trips to Paris, the Gauberts arranged for him to have a chaperoned "date" with the daughter of the president of France, who Clayton said was living with the Gauberts during the war. Clayton may have misunderstood the family background of the young woman, whose name he couldn't recall. The president of the French Republic from 1913 to 1920 was Raymond Poincaré, and he and his wife were childless. Of course, Clayton's date might have been with the daughter of another high government official or even a past president of the republic.

The date was a one-time event, and, he said, "All I did was reap fond memories."[7]

Clayton's relationship with the Gauberts was close (he called them "good people") and continued after the war. Suzanne Sykes-Gaubert visited the Williams family in Texas in 1924 and invested several thousand dollars in an unsuccessful business venture organized by Clayton and Waldo to grow cotton in Mexico. Janet Pollard recalls that "when the Mexican venture failed, my father paid back to the Gauberts the money they had invested." Other stories also circulated about the visit. "At the time," Janet says, "there was some concern on my grandparents' part that Madame Sykes's interest in my father was more than just friendly." O. W. mentioned the visit in several letters to Clayton's brother J. C. and in one commented, "I am afraid that [Suzanne Sykes] is planning to marry Clayton and try the second American husband." Clayton also believed that Suzanne Sykes-Gaubert's interest in him was more than platonic. "Oh yea, I'm sure Mrs. Sykes would have come over here and wanted to marry me, but I didn't want a family already made. And I wasn't in love with her."[8]

At the request of Madame Sykes-Gaubert, Clayton wrote his father in March 1918 seeking information concerning how she could obtain a divorce from her American husband and whether it would be quicker to obtain the divorce in the United States or in France. "She has asked me to write you and get your opinion in the matter," Clayton told O. W. "I solemnly told her you never practiced [law] on divorce matters and I could as easily get the address of some N.Y. lawyer who practiced on such cases for her, but her wish remained the same; so through courtesy to her and her family I am writing you as much of the details as I know."

Those details included that Madame Sykes-Gaubert believed it would take two years or until after the end of the war to obtain the divorce in France; that her grounds for divorce were desertion and mistreatment while she was "in a family way"; that her husband was drunk often; that her husband had left her in France with no money and then ordered her to come to the United States without providing her with any money for her travel; and that she believed her husband "thinks something of the steneographer" to whom he had mailed his letters, apparently for translation.

"This is no matter of mine," Clayton concluded, "only I feel somewhat indebted to the family and what you can advise there for their good or interest will be a courtesy to me as well as to them." No record could be found concerning O. W.'s reply to this request, and the matter of the divorce is not mentioned again in Clayton's war correspondence.[9]

Fort de la Bonnelle

Fort de la Bonnelle, Clayton's next duty station, had been constructed in two stages from 1869 to 1885 as part of a defensive system of forts built by the French to protect the militarily strategic Plateau of Langres in eastern France. The fort had been abandoned for some time prior to the outbreak of World War I, and upon his arrival Clayton found it a literal pigpen, inhabited by hogs and pigs in the underground rooms. During his first few days there, Clayton and his fellow officers had to transform the old fort "into a sanitary camp for the winter by cleaning, burning, draining and adding Swiss huts for the kitchen and dining rooms."[10]

Clayton's first formal training at Fort de la Bonnelle lasted from December 3 to December 17, 1917. Clayton, along with five other officers, was assigned as an instructor and taught officers and men of Maryland's 117th Trench Mortar Battery of the U.S. Army's 42nd Division, also known as the "Rainbow" Division. The division, composed of National Guard units representing twenty-six states and the District of Columbia, was commanded by Douglas MacArthur at the war's end. The nickname was allegedly bestowed by MacArthur, who said that because the division was composed of men from so many states, it "would stretch across the whole country like a rainbow." The division saw its first action of the war in February 1918, fighting alongside the French. Clayton, in addition to his instructional duties, was given a special assignment to design, procure materials for, and install the camp's electrical plant and lighting system, and he participated in the restoration of trenches for both defensive and training purposes.[11]

Christmas at Fort de la Bonnelle was a bittersweet affair for Clayton. He wrote home a detailed description of the day, noting that while he had a pleasant time, "the real spirit of Xmas was lacking. We lacked our homes, loved ones, dances, banquets, etc." The highlight of the day was a bountiful dinner, but even that was slightly off for Clayton as he and his comrades were joined by a number of other officers "more for diplomacy than congeniality." However, the large number of outside officers attending ensured a successful meal. "In this way," wrote Clayton, "when we want supplies, it is presumed that we shall have first attention."[12]

The guest list produced the desired result. The Swiss hut that served as Clayton's officers' mess was adorned with candles and other Christmas decorations, and the table was spread with a cloth. The menu for the 2 P.M. meal consisted of figs, olives, dates, oyster stew, roast turkey, potatoes, cranberries, brussels sprouts, tomato rice, apple pie, cherry pie, Roquefort and Swiss cheeses, cognac, Benedictine brandy, champagne, fruit, candies, and nuts. Cigars followed the meal and music was provided by the company band, which consisted of a piano, two violins, a flute, a drum ("consisting of a bucket, a pan, a chair, two sticks and a sheet metal"), a guitar, and a mandolin. After dinner, the group retired to the clubroom, which had just undergone the transition from pigsty. The afternoon was spent in "conversing, reading the papers (English) from Paris and 'musicing.' You see, it wasn't bad," Clayton wrote, "but still there is no place like home on Xmas day."[13]

At the Front

Clayton frequently failed to provide his precise geographic location in his letters, no doubt due to the demands of military censors. According to one expert on military censorship, Myron Fox, World War I marked the first significant censorship of American soldiers' letters. Letters were not subject to censorship prior to the American Civil War, principally because "most of the troops before then were illiterate and officers were trusted." During the Civil War, letters that crossed enemy lines, primarily from prisoner of war camps, were censored to remove descriptions of the harsh camp conditions. In both World Wars I and II, censors were concerned about letters that mentioned troop locations and the strength of military units, and they also looked for content that could be interpreted as a weakening of troop morale. Letters written home in languages other than English were confiscated, as were letters that contained graphic sexual language.[14]

"In some places my descriptions have been necessarily scant and brief for you know we are under orders about some features of our life on the way and here," Clayton wrote in 1917. To emphasize his concern that he follow censorship rules, Clayton said he believed the explosion in Halifax harbor might have been caused by a German spy who benefited from information that slipped by the censors. "I suspect that somehow some spy did get there and he may have formed his plans . . . because of the indiscreet letter of some loyal soldier, which gave him the necessary information."[15] Later in the war he wrote, "When I try to make it interesting to you, I find myself continually running up against the confounded Censors' regulations. I am always trying as I write to work out as a puzzle whether this or that item will be against the rules. Under our earliest regulations a letter could only say in substance, 'Hello, I am well, Goodbye.' But now we are let off with warning not to write anything which might convey information of troop locations or movements, or anything of military value to the enemy."[16]

Shortly after New Year's Day 1918, Clayton was ordered to attend a trench-mortar training school conducted by the British Second Army in the Ypres sector of Flanders. Clayton attended the school from January 6 to January 19 at St. Omer, just south of the English Channel ports of Calais and Dunkerque in northern France. At this stage of the war, American officers were routinely sent to British schools for instruction in the six-inch Newton mortar, a medium-range mortar similar to the French 58-mm No. 2 mortar.[17]

Graham Pollard, Clayton's grandson, says that his grandfather once told him that at some point during this time in France he received a special letter. Graham recalls: "I was 10 years old and visiting in Fort Stockton. I was watching a Bugs Bunny cartoon on television with my grandfather. Bugs was portraying George Washington during the revolutionary war and opened an envelope from the mailbox. He shrieked, 'I've been drafted.' My grandfather laughed and said, 'I was hip deep in a trench in France when I received a letter that had followed me halfway around the world. It was my draft notice.'"[18]

During his time at the British school, Clayton learned that some of the British officers were dissatisfied with the war's progress. "In our intercourse with these British Officers [at the Second Army school], they frankly admitted that we were 'keener' than they on the war and the training for it, and they gave us as a cause that they were 'bloody well fed up' on it." Statements of dissatisfaction about the progress of the war were common from the tired British and French officers and troops throughout all sectors of the fighting. The British Second Army had been charged with helping repulse the 1914 German advance through Belgium into northern France, and that effort evolved into a stalemate that remained until the end of the war, with gains and losses by both the Allies and the Germans occasionally measured as a few miles but more frequently in hundreds of yards. Gen. John J. Pershing, AEF commander, was worried that the French and British officers training troops were psychologically tired and demoralized and that their pessimism would dull the aggressiveness of the Americans. Also, during the spring and summer of 1917, soldiers of almost one-half the French army refused to take part in fresh attacks until their living condition improved.[19]

In a February 4 letter, Clayton said he had to "fess up" to his parents that he had been hospitalized with scabies after returning from Flanders and had been in a military hospital for ten days. "For one month, I did the best I could under conditions in Belgium, but had to get to a hospital as soon as I got back for some kind of 'relief.'" Scabies is a parasitic skin condition caused by unsanitary living conditions; it was a common ailment among soldiers during the war. Clayton believed that he had caught the disease through his underclothes. "The French peasants are not cleanly and rarely boil their washing. As a result, I and my comrade had our washing done together and changed underclothes at the same time; (we) noticed our itch started at the same time from those underclothes."[20]

Clayton experienced his first real taste of trench warfare during his time in Flanders. He described this experience in a revised serial letter written after the war, adding details that could not be included in the original letter due to the constraints of censorship.[21] As part of his training was to get actual experience at the front, Clayton and three other American officers left St. Omer on January 8 in the charge of a British captain. They traveled east by truck train through the town of Steenvoorde to the village of Bickebusch, from which they walked to the front lines. Clayton described the last few miles:

> At this point we were in cannon range. The long range guns and the aeroplane bombs had wrecked the village of Bickebusch, and here the continuous roar of the big guns on the front was painfully loud. Yet it was a good 6 miles away to that front. We walked this distance to the South part of the Ypres sector just north of Messines, famous for the fight on that ridge, and that walk will long remain in my memory, as it was my first real exposure on a firing line.
>
> All along the line of this very long walk of ours the big shells were dropping. At my first experience with a big shell-burst near me, my hair fairly bristled, and while I was never a star at the high jump I felt that I could easily right there

out-do any of my College records in that line. We were afforded very little protection other than an occasional house,
small wood or hill and when a bursting shell tore up the earth
near us it was almost impossible, at first anyhow, to avoid making a high jump. In time I got so that I could do away with that
jump, but, as there was many another American boy over there
who went through the same test, it will not do for me to say
that I entirely got over that fright. About all I can say is that
after a time I got so I could perform all my acts normally under
a continual sense of great uneasiness."[22]

Clayton wrote that he and his comrades were billeted in a dugout about one
and a half miles behind the front lines and were assigned to a trench mortar
station about 1,700 meters (about 1,860 yards) from the German lines. Here
they took turns firing at different targets such as German trenches and wire
entanglements. One day Clayton was assigned to direct mortar fire at an old
brick house in no-man's-land between the two armies:

> So I went up into the front line Observation Post with my
> field glasses, range table and pad. The O. P. was connected by
> telephone with the battery, so I directed the fire from there. . . .
> After judging the distance from the battery to the house to be
> 1600 yards, estimating the deflection from the map, and us-
> ing the range table to decide the angle of elevation, I directed
> the shot. After a few seconds of breathless waiting on my part,
> I saw a large up-burst of earth to the right of the old house,
> about 1 degree as I thought, and about 50 yards away from the
> ruins. I waited a few moments to be certain that the mortar
> was ready before giving a new setting, which shortened the
> range 100 yards and put the deflection at 2 degrees to the left.
> This time the shot fell short about 75 yards and to the left
> about 1 dog [degree]. So I was fairly certain now that neither
> of the shots had been abnormal. Bad powder or something else
> caused the next shell to burst not far from us, and believe me,
> I got down below my little observation slit in the trenches in a
> hurry, for the fragments might have come through on me.
>   But the next shot with the same setting had better luck, for
> the bomb hit the old ruin and up came a mass of flying bricks,
> mud and debris of all kinds showing out of a haze of smoke.
> Among the wreckage thus sent up, I thought I recognized
> some gray pants high in the air, but whether or not they carried

legs in them I did not know. But at any rate I felt sure that I had succeeded in separating a Boche from his trousers. As no one was ordered to fire again on the old house, I suppose it was considered that I had done my task thoroughly.[23]

Because of censorship, Clayton was much more circumspect about his time at the front in his hand-written letters home than he was in the serial letters

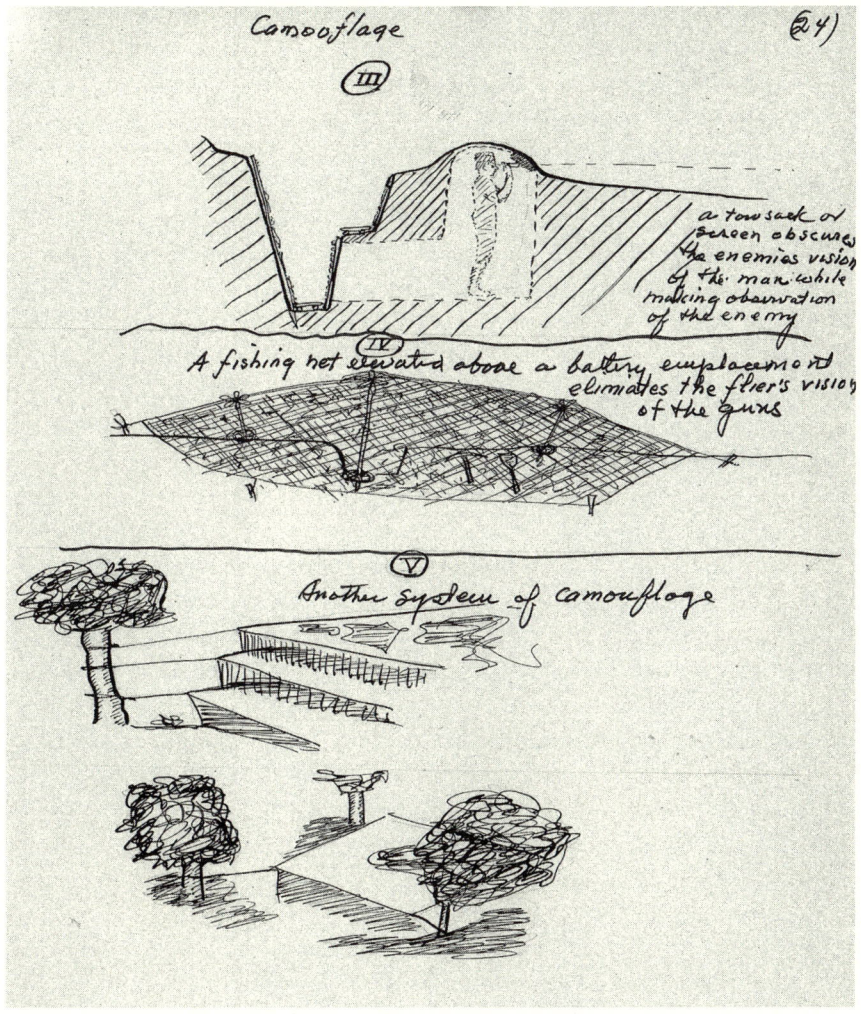

Drawings by Clayton showing methods of camouflaging artillery installations from enemy aircraft. Clayton spent part of his time during the war training other American soldiers in trench and field artillery. He used drawings to illustrate an informal instructional manual on trench mortar warfare.

he revised after the war. In a letter dated January 27, he simply wrote, "I've seen the flash of the big guns and heard their continual roar; the airplanes drop their signal torches to the captive [observation] balloons way high up." "Captive" balloons were tethered to the ground by one or more cables and were winched down quickly when enemy aircraft approached to try to shoot them down. Observers in the bucket below a balloon, on a day with good visibility, could see as far away as sixty miles and directed artillery fire using a one-way telephone line attached to a tethering cable. If the balloon was shot down, the observers leaped from the bucket and parachuted to the ground, often a risky proposition.[24]

In a passage concluding that letter, Clayton vividly described a war scene that he characterized as a "dream," writing after the war that he had called the scene a dream to "pass the Censor":

> I was walking back about 3,000 yards from the first live
> trenches, and high above me were two captive balloons. Sud-
> denly an aeroplane seemed to dart a wing into the nearest cap-
> tive, and immediately afterwards the three men jumped out
> from the stricken balloon. Two of them successfully opened
> up their parachutes, but the third failed. The unlucky man in
> falling grasped at, and for a time clung to, the cable of the bal-
> loon, but after a few seconds lost it and plunged straight to
> the ground some 400 feet. A moment or two later there was a
> thud behind me, and looking up I saw that the 'Sausage' had
> been split and was collapsing. Running back toward the place
> of the thud, as it seemed to me, I found an old well filled with
> water to within 8 feet of the top, and in the muddy water I saw
> an elbow sticking out of the slime. Calling [for] help, I crawled
> down into this well, and got out a man, who was able to groan
> and no more. We lifted him out and managed to pour a little
> brandy down him, which gave us some hope he would live.
> Later I saw him in a peasant's hut, living yet, but badly hurt,
> and after that I heard no more about him whether he made it
> through or passed to his account.[25]

Later in his life, Clayton recalled a more amusing incident that happened during his time at the front when a Canadian soldier offered the young Texan a drink of cognac. "Well, my feet were wet. It had been drizzling ... and those trenches were terrible, and I said 'yea.' He [the Canadian] took one himself; he really took a swig of it. I tried to do like he did and it liked to knock me over. I just turned my back to him with tears running out of my eyes and handed the bottle back to him."[26] Clayton said this was his first experience with cognac, although he had the opportunity to sample the spirit both during his first visit with the Gauberts and at the Christmas dinner at Fort de la Bonnelle.

Another story that stayed with Clayton demonstrates the unimaginable horrors of trench warfare. He was told that a soldier had been killed by a sharpshooter while walking down a shallow trench leading to a latrine. "They dug in the side of the trench and buried him right there. If I remember correctly, his hands deteriorated and were still sticking out of the wall."[27] Many years later, Clayton left the theater during a showing of the movie *Patton*, telling his grandson Clay Pollard that he had been in war and that he didn't like to see the suffering on screen again.[28] Grandson Graham Pollard says that once when he was about six years old his grandfather had bought him a kazoo in the shape of a bugle. "I played songs on the kazoo that I thought were appropriate for the bugle, including 'Taps.' After I finished, I noticed my grandfather had been noticeably moved by my playing. A couple of decades later I thought back to that day, and I understand what 'Taps' means to a soldier."[29]

While censorship prevented Clayton from telling the complete story of his trip to the front, he recognized that his family would be concerned about his proximity to the fighting. In the letter of January 27, he apologized for not having written in three weeks. "I am sure you will pardon the delay," he wrote, "and will feel that while I am the same old Clayton, growing wiser, I hope, from experience in as dry a climate as West Texas, [this] has not equipped me to meet the many inconveniences of winter mud in Flanders." Later in the same letter, he wrote that he had "arranged so that if anything happens to me, you will hear about it regardless of the official announcement, and of course if it doesn't happen you will not hear."[30]

Anticipating that he would be exposed to enemy fire in Flanders, Clayton wrote his father on December 23, two weeks before attending the British Second Army School, that he had applied for a $10,000 insurance policy, noting that "it never rains when you carry your umbrella and slicker." Clayton told his father that should he be "totally disabled—two legs, eyes or etc.—I am due (or you are due) six months' pay in case of death besides insurance." He also urged his father to keep his younger brother J. C. from entering the war. J. C. tried to qualify for service as an aviator but was turned down due to poor eyesight. He was commissioned as a second lieutenant of infantry in September 1918 but remained in the United States as an instructor throughout the war. During World War II, because of his experience in China, he served as a colonel and chief intelligence officer for Maj. Gen. Claire Lee Chennault's Flying Tigers. Clayton's older brother

Waldo attempted to enlist in World War I, but was turned down for health reasons.[31]

Clayton also strongly discouraged his sixty-four-year-old father from traveling to Europe, a suggestion raised by O. W. that perhaps reflected the father's concern for his son's safety. While there are no copies of O. W.'s letters to Clayton written during the war, a number of letters written by O. W. to his other children during the period have been preserved in the Haley Memorial Library & History Center and also in the University of Texas' Dolph Briscoe Center for American History in Austin. In these letters O. W. described Clayton's activities in the war for the benefit of his other children and occasionally expressed his concern for Clayton's safety. For example, in letters to J. C. dated July 23 and July 25, 1918, O. W. noted that he had not received a letter from Clayton in several weeks but had worried unnecessarily as it turned out that Clayton had been sent to fight in what would later be called the Second Battle of the Marne.[32]

It's Not So Great "Over There"

Despite the censors, Clayton was not hesitant to criticize the conduct of the war and the way it was reported in the American press. One particularly harsh critique was written two days before Christmas 1917 and perhaps reflected Clayton's low spirits at being so far from home during the holiday season, and also his concern that his letters home (twenty-five by his count) were not getting through. His family had sent him newspaper clippings and magazines that contained war coverage. "These I read with great interest for I can compare [them] with the news which I receive. The news you [his family] get has been colored and shaped around to content the public. You hear all the small minute victories which the allies make, and their big and clumsy defeats are covered by details of their valiant resistance against overwhelming odds."[33]

In particular, Clayton cited the disastrous Italian defeat and retreat at Caporetto in northern Italy in October and November of 1917, and the Battle of Cambrai, in which the British launched a successful attack with tanks on November 20 along the western front in north central France, only to lose much of the captured ground in a German counterattack ten days later. "Hear-say is that the preparations for a huge attack by the Germans on the Italian front were secretly known," Clayton wrote. "The French and English offered to send reinforcements to help but [Italy] confidently refused. This is just a general misunderstanding over here. You have seen what happened. Again, the English victory at Cambrai? Ah! What a sad mistake. They advanced easily only to be donkey tricked and beaten. 'Tis considered true that one company of American engineers fought nobly here when surprised by the Germans and surrounded by barrages. Well, anyway, the papers came out with the statement that the lines had been straightened for tactical reasons."[34]

Clayton was also disheartened by the lack of American preparedness for the war. Commenting on his effort to buy supplies in New York before leaving for Halifax and Europe, he noted that "it came about that the commercial people were not prepared to supply all of us [officers], after 3 months notice. The price rose at once. The result is that many commissioned officers are over here now, not equipped in pistol and holsters, ordnance, blankets, and other things. Many a man has been saved from pneumonia by the large and heavy blankets furnished by the French and English."

He called articles in magazines and newspapers praising individuals and concerns supplying the war effort "paper rot" and told his father that none of the preparations "are sufficient in either vastness or detail." He noted, for example, that there had been four cases of meningitis in one of the army hospitals in France and that "three are dead and the fourth is dying simply because they have no serum. Pneumonia is nearly as deadly for the same reason." But he did reserve praise for Henry Ford, whose "little jitneys save many a man's life by taking him quickly from anywhere over anything to the Hospital."[35]

Clayton's complaints reflected the fact that the winter of 1917–18 was the most severe of the war, and clothing and boots were in short supply for the American troops. However, the army's ability to train, supply, and transport soldiers improved significantly after the appointment of Maj. Gen. Peyton C. March as army chief of staff in March 1918. Between April 1917 and December 1919, the AEF lost 58,119 men to disease, with pneumonia accounting for more than 80 percent of the deaths. Total American war deaths were more than 116,000. While estimates vary, American casualties during World War I were more than 250,000.[36]

Tractor Artillery School

Clayton was hospitalized with scabies from January 23 until February 7 at Base Hospital No. 15, a 3,000-bed hospital organized by the American Red Cross in early July 1917, at Chaumont in the Department Haute-Marne, located west of Paris and north of Dijon. He engineered his release after learning from a fellow hospitalized officer that some of his comrades had been sent to school to study autos and motors, and Clayton suspected that he had also received similar orders. A telephone call to Fort de la Bonnelle confirmed that he was assigned to a school for tractor artillery at Vincennes near Paris and that he was several days late. Clayton camouflaged his illness by applying ointments "so that I was successful in my plea to get away." He returned to Fort de la Bonnelle the next day and "gathered up as much of my plunder"—by which he meant baggage—"as I could, and left for Paris." He reported on February 10, having missed a substantial part of the four-week course.[37]

Clayton was sent to the school because the U.S. Army was making the transition from horse-drawn to mechanized artillery and needed officers to learn the new technology. The development of the Caterpillar tractor and its use in agriculture had convinced the military that it could be used to haul heavy loads over soft ground under conditions that would have been impossible for horses. By 1914, the British, French, and Russian governments had all ordered Caterpillar tractors manufactured by the Holt Manufacturing Company of Stockton, California, for use in materials handling, supply movement, and artillery traction. Though the U.S. Army was slower than its French and British counterparts to adopt motorized transport, it began using Caterpillar tractors to move guns and ammunition in 1916, and by the end of the war Holt had produced nearly 3,000 tractors for the U.S. military.[38]

Even though tractor-pulled artillery had limited use during the war, it was clear that mechanized artillery units could move across difficult terrain more quickly than horse-drawn units. Moving horses to the battlefield required more shipping tonnage and freight cars than moving mechanized units, and the supply of horses and forage was limited. One senior artillery officer summed up the use of horses in artillery units thusly: "Personally, I think to serve with horses is very pleasant and I am very fond of them, so I should say that from the standpoint of sentiment they are greatly to be desired, but from the standpoint of utility in the Artillery they are doomed in favor of motor equipment."[39]

Clayton described his abbreviated two-week course as a "somewhat technical" one that involved learning the basics of automobile, truck, and tractor operation, as well as maintenance and repair for a number of different engines, including the French-made Latil and Renault tractors, the four-wheel drive Jeffrey Quad truck, and the Packard truck. "Our study was especially directed to automobiles," Clayton wrote home. "We went into it thoroughly, from magnetos, cylinders, and pistons, to the differential, and our course was as thorough as could be had in the time." Clayton also learned how to use tractors to move and position the French-made 155-mm Grande Puissance Filloux (GPF), the principal long-range artillery piece used by the AEF during the war.[40]

During one training exercise, Clayton and his fellow students were required to use tractors to move four howitzers some 15 kilometers (more than nine miles) into place on a steep, sandy hill. It took twenty men per tractor to do the job. "It was snowing and raining, and it was thus a test under difficulties," Clayton wrote his father. "We started out at 7 A.M. and pulled out the 15 kilometers, put the guns in place, which latter task required some road building and work with blocks and tackles among the big trees and sand drives. [We then] pulled them out again and got back by 10:30 P.M. Now it was good work to pull 4 guns, each weighing 15 tons, out and back 30 kilometers, place them in and out of position under such difficulties, all done by tractors handled by us alone." Always convinced of American military superiority, Clayton commented that "the French were so taken with the feat that they have been proud to work personally as we did in this. But the American idea over here is to do it yourself, then show others how."[41]

At the end of the course, Clayton took a number of difficult oral and written examinations. He wrote that of the 150 men taking the course, thirty-five failed and were sent back to their respective units. Some twenty officers, including Clayton, were asked to repeat the course, while the rest were sent off to be instructors. Perhaps to make himself feel a little better, Clayton wrote home on March 11 that "the Colonel [Lt. Col. Clifford Carson, AEF director of Tractor Artillery Schools] says these men should feel highly complimented, for they are supposed to make *excellent* instructors by a little rounding off and that some of the highest [scoring] men were repeating it." He also told his family that his report from the British Second Army trench-mortar school had been "*very good.*"[42]

Back in the Hospital

Clayton's exertions in the wet and cold of the French countryside eventually got the better of him. After only one week of instruction in the tractor artillery course, he found himself back in the hospital, this time with symptoms of "pneumonia or pleurisy." He did not say when he entered the hospital, but in a somewhat disjointed section of a letter dated March 11, he complained of chills and fever, and by March 15 he was in American Red Cross Military Hospital No. 3 in Paris, where he remained until his release on March 23.[43] His illness was apparently not serious, and the hospital stay afforded him time to write some long letters home containing lengthy observations about the war effort, the French people, and France in general.

His hospital stay also marked the return of Suzanne Sykes-Gaubert and her mother to Clayton's war correspondence. He had stayed in touch with the family since his first visit in November 1917, and the mother and daughter arrived to visit him on March 15, shortly after he was hospitalized. They reached his hospital bed just seconds after a large explosion destroyed a hand-grenade and munitions depot in La Courneuve on the northwest outskirts of Paris.

Clayton wrote home that he was looking out his window and "heard a loud explosion, followed by a louder and then by a weaker one—something in sound like that of bombs—but surely the Boche would not have the gall to bring their Gothas over Paris in broad daylight. After a few minutes, I saw a large white cloud rising in a dome-like form across my blue sky horizon, reminding me of our 'white caps' when the home summer rains start. Under this, and seeming to push it up, there was a central stem of smoke in great quantity, moving with the apparent force of a tornado."[44]

Suzanne Sykes-Gaubert and her mother were at first concerned that the Gaubert factory had exploded but soon learned in a telephone call from a relative that the factory had survived, suffering only some broken windows. Accounts of the blast indicated that windows were broken several miles from the center of the explosion. One description written by an American living in Paris at the time noted that windows were blown out along the avenue de l'Opera and that closer to the explosion walls were cracked and houses destroyed.[45] Sharing Clayton's views about American superiority, the observer noted with satisfaction that "the American Red Cross and the YMCA ambulances were on the spot within twenty minutes of the explosion and our men did splendid work. The French never seem to get used to our national characteristic of quick initiative. With them, there is always so much formality and red tape before anything can be accomplished—even emergency aid to the injured."[46]

The *New York Times* reported that the death toll was initially sixteen, but the next day raised it to thirty, and that a thousand people were left homeless by the explosion. Many Parisians initially feared that the depot had been destroyed by German bombing, for the city had been attacked by bombers twice just a few days earlier. However, news reports at the time indicated that the explosion had been accidentally triggered by a worker in the depot.[47]

CHAPTER 5

Paris Is Bombed

Germany began making night bombing raids on Paris in 1917, and a few days before the La Courneuve depot explosion, Clayton witnessed two nighttime air attacks on the city. In a March 8 attack, thirteen civilians were killed and fifty were injured, while on the night of March 11 some sixty German aircraft dropped bombs that directly killed thirty-four civilians and injured seventy-nine. Another sixty-six civilians, mostly women and children, suffocated while taking shelter in an underground subway station. U.S. Sec. of War Newton Baker was in a meeting with American generals in a Paris hotel when the raid took place and took shelter in the hotel's wine cellar; he was not injured.[1]

In a long letter of March 19 to his father, Clayton vividly described both attacks. "The constant discharge of guns [on March 8], followed by the explosion and dispersion of shrapnel high up in [the] sky, visible only by the flashes among the stars, made up a highly exciting scene. Far off rockets exploded and, letting out parachute lamps, lighted up the horizon as they slowly descended with an ever lessening blaze. . . . Up high in the sky once, and only once, I saw a machine gun spitting fire down on some object below, and evidence . . . of a French machine fighting the raider. I suspect it was the only one able to get out into action."[2]

The March 11 bombing struck much closer to home when a German bomber flew low over Clayton's officers' quarters.

> Just after it passed us it dropped a bomb, and as it did so, all the
> anti-aircraft artillery in that neighborhood shut up shop and
> kept quiet. Then it circled slowly in the air, and came back over
> us for a second time, and it began to look like we were their
> special mark. A poor fellow next to me was shaking as if he
> had a very bad case of chills. He had some cause of his feeling
> for we began to think we were the only mark clearly visible to

plane, and we were in for it. But its bomb again was dropped just beyond us, landing in Fontenay. After this it took its course away from us and towards Fort Vincennes . . . and then came back again. Rumor has it that there is a very large ammunition plant not far from us, and if so then the plane must have been trying to locate it. If it should have succeeded in planting a bomb there, then we would have fared badly. . . . This time it dropped a bomb 300 or 400 yards before getting over us, and then another which struck about 50 yards from our gate. The last wrecked two houses, but killed no one, only took the roof off of a truck. On this parting shot, it went away.[3]

Concluding his eyewitness account of the two raids, Clayton critiqued the French on their response.

On this night, I saw no evidence whatever of French planes, and the Gotha seemed to have the field to its own use, undisturbed. It is possible that the raid of the previous Friday night had put the French plane[s] out of commission. . . . It was stated by some Frenchman on the following day that the bombs were dropped by the Gotha because it was pressed so hard by a French plane that it had to lighten its load in order to get away. You will know, from what I have told you I saw, that this is French 'hot-air.' Peculiarly, it seems to me, the French are given to making assertions of facts which they really surmise or suspect only. I am only entitled however in my turn to say that my conclusion is of course only drawn from my experience with a few of the French officers, and may be unjust to the nation at large. On the contrary I have found among the British army officers with whom I have been thrown, that an Englishman will admit he doesn't know, when he might have a strong suspicion only.[4]

Clayton had mixed feelings about Paris and the French people. To the Parisian, wrote Clayton, "Paris is 'France,' recognized by the world as its greatest city, whose women are such that they outshine all other women in the grace with which they carry the gems and adornment of dress." In reality, he believed, Paris "is a city of several million inhabitants, every one holding both hands out for a tip." He acknowledged the beauty of the Champs Elysées, Arc de Triomphe, Paris Opera House, Place de la Concorde, and of Napoleon's monument and Notre Dame, but he was also critical of these expensive memorials:

> Over these luxuries they have erected structures of immense cost to protect them [monuments] from the bombs of the Boche, while not a structure has been raised as a house of refuge for the passerby during the same danger. The public are let off with simple warnings to take to the underground 'Metro'— long ago built and not always accessible to every one, when the hum of the Goths is heard.
>
> . . . These monuments are to be preserved for the tourists after the war, and from them rich and poor will reap a harvest. Certainly that is what we Americans call a practical idea and the Parisian is surely practical along those lines. They are too practical it seems to me just now, for they are taking advantage of the average American's inexperience in France and his ignorance of the French language and customs to charge him exorbitant prices. Articles bought by a Frenchman cost about 1/3 as much as when bought by an American. This is prejudicing our soldiers against the French, and I have heard of one of them remarking that the 'French are making more money since the war than at any time before.' In fact our boys are betting their eyeteeth here, and will be careful bargainers when they get home.[5]

He described Paris as a "sanitarium so skillfully conducted that nobody can tell the patients from the physicians; and all the inmates are firmly convinced that the outside world is mad."[6] Later, while stationed outside Paris, Clayton observed that "time has not changed France much since the days of those old Cathedral builders, save only in spots which have been selected for the display of some kind of scenic beauty. The houses in general are old, breezy and cold, with few fireplaces, no bathrooms and no accessories for convenience or comfort."[7]

This observation sparked a story from Clayton about an unnamed American army officer who needed a bath. Either unable or unwilling to locate an indoor bathtub, the officer found a watering trough, removed his clothing, and began bathing in the open air. "It was a rare scene and the French did not miss it. 'An American bathing out in a water trough in plain sight' soon passed around, and there was a large crowd of French looking on in amused attention. The man had his nerve evidently, and he proceeded calmly and deliberately in the face of the crowd and its laughing jabbering spectators.

The only penalty he paid was a trip to the Hospital under the attention of LaGrippe [influenza]."[8]

Clayton believed that the French people were in better financial shape than their counterparts in England because of money the British and American armies were spending in France. "Many of the French, yet living, are making more money now than at any time heretofore. . . . So while France has suffered much in men and money, affairs are not so bad now with the people, and it must help their morale. To be sure, the little street urchins strike one everywhere for a donation, but this habit must be from old, and Americans—as reputed millionaires—are their especial prey. Really I suspect there are more downright poor people in America than France, but as my judgment is that of a casual observer, it is not worth much stress."[9]

In an aside, he expressed his astonishment at the French love of wine. "Actually, wine is more plentiful here than water, and is probably substituted for it . . . entirely by many French people. You know this, and I had heard it before I came to France, but I was just a little skeptical about it. Now I am learning better. It may be more in evidence these days than usual because nearly all the eatable luxuries of life are now under quarantine and are seldom seen."[10]

Germany's 1918 Spring Offensive

Clayton was released from Military Hospital No. 3 on March 23, just as the Germans were preparing to launch a major offensive along the Western Front. The war in the east against Russia had effectively ended with the signing of the Treaty of Brest-Litovsk on March 3, 1918, allowing Germany to concentrate its forces in the west. In a letter written on March 28, Clayton noted that on the day he left the hospital and for two days after, Paris was bombed by a German long-range gun. This was not, as Clayton believed, the German 420-mm heavy howitzer known as "Big Bertha" that destroyed Belgian defenses in 1914, but the experimental long-range guns known as Paris Guns that could fire shells into the city from as far away as 80 miles. The guns fired more than 300 shells into Paris between March and August of 1918, killing 256 and injuring more than 600.[11]

In the March 28 letter, Clayton described the major German offensive against the British Fifth Army near Saint-Quentin, which began on March 21 and over a period of several days advanced to twenty miles east of Amiens and the road to Paris. "The old Somme Line promised to be the scene of another terrific struggle," he wrote home, referring to the ill-fated 1916 British offensive in the same general area north of the Somme River. "American troops will soon be in this. Already American aviators (248) have been working in this battle." Indeed, American divisions moved into the trenches of the

French army sector to the east of the British Fifth Army in April, but Clayton's reference to American aviators is puzzling, since AEF fighter aircraft did not make significant contributions to the war effort until the summer of 1918. However, a small group of American pilots had fought with the French air corps in the Lafayette Escadrille since the beginning of the war.[12]

A keen observer of the progress of the war, Clayton knew that the German offensive was serious and of much concern to the Allies despite newspaper reports to the contrary. He noted that people were leaving Paris from fear of the German advance. Some 500,000 Parisians evacuated the city in the first days of the advance, and the Americans assembled a fleet of trucks in case it was necessary to evacuate the American embassy.[13] "The papers kept assuring the citizens that the Allies were holding the enemy, but when the positions of the Germans were reported daily, it could easily be seen that they advanced about ten miles a day for the first two days," Clayton wrote home on April 13. "The third day they did not do so well, and on the fourth day they were being checked. But they were more successful than most people will admit." In this letter, he credits the effective use of Allied artillery with halting the German advance, although military historians believe that the advance was halted for a variety of reasons.[14]

Perhaps due to the initial German success in the spring of 1918 and the fear that was engulfing Paris, Clayton conveyed to his parents his frustration at continuing to be far from home when the goal of winning the war was uncertain. As he often did during these "low" periods, he complained in an April 5 letter about the mail service to Europe, adding his displeasure at the lack of mail received from his family. "The letters I write can go around, but I do not get an answer for every one I write," he wrote his sister Kathryn. "Of course Father's letters are always dependable and I can tell you that this makes a big soft spot in my heart for him. J. C. hasn't dropped me a line since I left the states—Waldo is in the same class. Ermine dropped me two fine letters. You and Mother I shall credit with several." Clayton also confessed to Kathryn that his uncertain economic future would probably result in his continuing to stay unmarried. "My chances for employment after the war will not be encouraging although my present study and familiarity with automobiles may aid in this outlook. Electrical engineering in my line may be past history. I am in a bad mood for writing tonight as you notice—so please excuse."[15]

Completing his course in tractor artillery on April 7, Clayton proudly reported back to his family that he was the "next to the highest man in the grade of Mechanical Maneuvers," which he defined as "handling of G.P.F. [Grande Puissance Filloux] guns weighing 15 tons by tractors, capstans, blocks and tackles." He said he rated high in the theory of mechanical maneuvers "and was told unofficially that I was among the first ten in the theory and study of automobiles."[16]

He modestly credited his successful performance in mechanical maneuvers to the fourteen men he supervised who handled the guns and tractors. "I owe them quite a bit of thanks for their exceedingly good work with me, and beyond all I am very glad to have met and known such men." Among his men was Archibald MacLeish, the 1915 Yale graduate who would go on to win three Pulitzer prizes for poetry and drama and briefly serve as the Librarian of Congress under Pres. Franklin D. Roosevelt. Another was Reginal Candler Foster, a Harvard graduate who would later be awarded the French Croix de Guerre for bravery in executing reconnaissance missions.[17]

Clermont-Ferrand

At the conclusion of the course in tractor artillery, Clayton and eighty fellow officers were assigned on April 10 to AEF Organization and Training Center No. 3 at Clermont-Ferrand in the Auvergne region of south-central France. Because the center was still being organized, Clayton was briefly attached to Battery F of the 56th Artillery (CAC) Regiment before being appointed to the center's automobile instructional staff.[18] He celebrated his twenty-third birthday on April 15 by drilling the troops of Battery F for two hours in a cold rain. "Poor fellows," he wrote home, "their shoes were flabbily wet from four drills on previous rainy days. They had no change because their baggage had not reached them, and their only extra pair was already soaked in the continuous rain we have here. Sympathy is all that they can count on, for the exigency of this war requires that our soldiers shall be drilled and [I] got ten ready for action at the earliest possible moment. Rainy days in the trenches are ahead and they must be hardened and taught how to take care for themselves there."[19]

Evidently the first few days at Clermont-Ferrand were difficult for Clayton as well. He apologized at the end of his April 15 letter for its quality, noting that "we are not yet well settled in our new quarters and we suffer many inconveniences, so that now I am finding it impossible to write a decent letter while my feet are cold. I am looking out on some tall snow-clad mountains, whose heads reach up into the clouds, and I am passing sentence on them as probably responsible meteorologically for our cold wet days. But that does not warm my feet. However I think we shall soon be more comfortably quartered. Otherwise, I can hope to do little good in preparing lectures on carburetors, magnetos, cylinders, pistons, clutches and all that stuff."[20]

Clayton spent the remainder of the war at Clermont-Ferrand, teaching auto mechanics courses to officers and enlisted men. Organization and

Clayton's identification card for his duty at Organization and Training Center No. 3 at Clermont-Ferrand in south-central France. Here, until the end of the war in 1918, he trained nearly 300 men in automobile repair.

Training Center No. 3 was one of a number of schools hastily established in France to train the AEF troops arriving in Europe. The fact that Clayton would not see combat during the war, with the exception of his brief time at the British Second Army school in Flanders, was not unusual. Due to advances in technology and the logistical requirements of a large army, for the first time in U.S. history more soldiers served as skilled technicians and in supply, transportation, and service roles than in combat roles.[21]

Between April 17 and May 2, while waiting for training supplies to arrive at Center No. 3, Clayton had time to write home five long letters that resemble abbreviated instructional manuals on trench mortars. He also included some drawings to illustrate various points in the text. "I am sending only the simple items and even these will be very interesting to the average reader," Clayton wrote. "Beforehand, I shall humbly beg to be excused by the womenfolk of our family, for bringing such a dry subject before their notice, but I hope it shall enlighten their viewpoint on the present war and give them a little satisfaction."[22]

In the ensuing letters, he describes how and where trench mortars should be sited; the proper location of observation posts to observe the effects of firing; the use of camouflage to conceal the mortars from airplane observation; and the use of airplane and balloon observation to direct accurate trench mortar fire. The letters are detailed and reflect the knowledge of a young man who had devoted a year of his life to the study of the subject. It is not clear from the letters why Clayton wanted to record this material. Perhaps he planned to publish a book on trench mortars at a later date and wanted to have a written record while it was still fresh in his mind. Or perhaps he thought everyone in his family would be just as interested in the subject as he was. As evidence of Clayton's enthusiasm for trench mortars, he wrote to his father later that summer that he was working on the design of a new type of trench mortar that could be transported by motorcycle but that he was too busy with his teaching duties to finish. "This would be quite an advantage in offensive warfare," he remarked.[23]

Return to Paris

With the arrival of the necessary training materials, Clayton plunged with gusto into his duties as an instructor at Clermont-Ferrand. On May 8 he was sent to Paris with three other officers to tour the Renault and Latil automobile and airplane engine factories. He took a night train to Paris, a trip of eight hours that he said "may seem slow traveling to you, but . . . is exceedingly good time to make these days in France." He attributed the delay to the baggage handlers whom he termed "slow drowsy old men" who "never get the bundles and packages on the [train] in any decent time"; everywhere "these steel rimmed and rail riding stage 'Busses' are held up from 10 to 20 minutes." Also, Clayton found the train cold and uncomfortable, and he slept with his clothes on.[24]

Upon arriving in Paris early on the morning of May 9, Clayton and his fellow officers signed in with the provost marshal, obtained tickets for bread, and checked into their rooms at the Grand Hotel, which was located at 12 boulevard des Capucines. After a shave and a bite to eat, they took the Paris metro to the Renault factory, where Clayton was singularly unimpressed with the French security arrangements. "We entered, passing the guard without presenting a pass or saying a word," he wrote. "This seems to us carelessness, as a spy could easily dress up as a French or American officer, and you may have heard that many men for a time masqueraded in Paris as American officers."[25]

What did impress Clayton at the Renault factory were the small, fast tanks being assembled by the French, and in his letter of August 12 he describes his vision of how the future of warfare would revolve around the tank. "The more I see of this war the more I am convinced tanks and cavalry in a large quantity and fast, with infantry to follow up, can win the war speedily, because these two facilities can put the large cumbersome artillery, which is effective now, at a disadvantage, and also they can follow up a victory with sufficient rapidity to shatter the enemy forces. Of course such could not be accomplished on a small scale nor in a short time, but eventually it would, I think, be quicker than the present methods." Clayton's view of tanks may also have been influenced by the successful British and French counterattack begun on August 8, during the Second Battle of the Marne, that used tanks and infantry to push the German army back to the original Hindenburg Line.[26]

While in Paris, Clayton had the extraordinary good luck to encounter his cousin Ned Holland while attending a movie on the evening of May 10. Holland had trained with Clayton at Fortress Monroe. "Chance threw us together at the same picture show at the same time," Clayton wrote his father. "I was very glad to see him and [to know] that he was doing well . . . on General Shipton's staff."[27] This was likely Brig. Gen. James A. Shipton, who was the first chief of the AEF's Antiaircraft Service and who commanded the army antiaircraft school and fixed air defenses in the rear areas.[28]

Auto Mechanics

Clayton wrote his father from his Paris hotel room during the May trip and said that he was one of four automobile instructors at Organization and Training Center No. 3 who had been selected to give lectures. "I may be lucky or unlucky there—depends upon what kind of a lecturer I make. Most certainly, I can't prepare anything but offhand lectures if I am on the go all the time. I was only at the center a week and had to correct [manuals on] magnetos, cooling systems and carburetors, besides adding drawings and descriptions of five American carburetors, and the starting and lighting system of a Dodge."[29]

At the end of the letter, perhaps anticipating that the war would go on for some time, Clayton told his father to have his brothers "go through my trunk and things. Put mothballs in and take care of them—what they or you can use you are welcome to. It will probably be three years before I even have a chance to use them." As spring moved into the early summer of 1918, more and more AEF troops were arriving in France for quick training before being moved to the front to help relieve British and French troops. By June 1918, the AEF was receiving 250,000 men a month in France and had twenty-five organized divisions at or behind the front.[30]

In a letter to O. W. dated June 9, Clayton told his father that his letter writing would have to be suspended for a while as he concentrated on his instructional duties. He also speculated that he might be called to the front. "I hear that all the men with previous trench mortar experience are being called in," he wrote. "Whether this will include me I do not know for I have a rather important place just now and I doubt that they will take me away; but if the call is urgent I may go."[31]

But the push was on to get the AEF troops trained and into battle, and Clayton was needed in Clermont-Ferrand. He wrote on June 23 that he had been "well occupied this last two weeks" as his mechanics' course had been cut to eight weeks, half of its original length. "So I have been working all day long and at night to prepare [students] and get out guide pamphlets for them in hunting down auto troubles."

Clayton's teaching job was made even more difficult by a lack of tools and vehicles. He had to go to his commanding officer to lay hands on two sets of carpenter tools for his men to build shop benches and boxes with.[32] "It was some job to keep 100 men with previous mechanical experience working and learning," he later wrote.[33] "I found early in the game the best and most effective method of doing things was to do it yourself and see it done right on the spot." The lack of English versions of French automotive manuals created more problems, which Clayton solved by finding English translations of the manuals in Paris during his trip to tour the automobile and aircraft engine plants.[34]

After successfully concluding his first course for auto mechanics, Clayton taught a second course that lasted six weeks and ended in early September. He was assisted in his teaching duties by several junior officers, including Clifford Herbert, the son of the famous late nineteenth- and early twentieth-century composer Victor Herbert.[35] He noted in his letter of July 20 that not all the officers had the personality to work with enlisted men.

"To get the men interested in their work, it is necessary to use men in charge of a proper turn, or manner, otherwise the work is a failure. On one duty or another, I kept one Lieutenant almost entirely away from the shop simply because of his disposition and manner toward the men." Clayton said that he knew every enlisted man in the course by name and that he kept a daily record of each man's work. "When the end of the course was nearing I knew every man who could time a magneto, time valve, etc."[36]

The hard work almost paid off materially for Clayton as he was recommended for promotion more than once and gained the respect of his commanding officer, Maj. Sidney H. Guthrie. However, he was never promoted during his time in the army, a fact that caused the young second lieutenant some bitterness. In his letter of December 10, 1918, to his father, Clayton noted that his latest recommendation for promotion did not go through despite his being the second on a promotion list of five officers. "I was the only one who did not get his promotion. . . . The Colonel telegraphed several times in regard to my promotion—but received no answer—well—there [is] something either rotten in Hq. A.E.F. or on my record!!!"[37]

Despite the lack of recognition in the form of a promotion, Clayton was proud of his work at the center and of his ability to "make do" in difficult situations. His courses in auto mechanics benefited the center in two ways. In addition to receiving instruction in auto mechanics, Clayton's students repaired automobiles that were used by the center to teach soldiers how to drive. According to Clayton, the center had 120 salvaged vehicles in its driving instructional program. "On these old worn out vehicles nearly 600 men were taught to drive each month," he wrote his father. "Well, you can imagine the amount of repairs to be made. . . . I turned my place [his classes] into both instruction and repair—making a regular outline and schedule on actual repairs. At the end of the 8 week course, the 100 or so men taking the course were good and I often turned out 20 cars a day." The AEF used many different makes and models of automobiles during the war, making Clayton's job even more difficult.[38]

Repairing the many and varied automobiles that came into his shop required Clayton to find spare parts. "I was the only one who knew what those parts were and on what cars they belonged—so it became necessary for me to make out a list of parts needed." Realizing that Clayton was the only officer under his command who could identify the parts necessary to keep the instructional automobiles operating, and also as a reward for Clayton's diligence and inventiveness, Guthrie sent him to Paris at least twice on parts-purchasing expeditions.

This required some ingenuity on Major Guthrie's part as his first request for permission to send Clayton to Paris was refused by headquarters on the assumption that the assignment should be handled by the army's supply organization. So Guthrie requested that Clayton be sent to Paris to "study operation of spare parts." This rationale was accepted. Clayton's first parts-hunting trip was on July 18. "I went, but not to study. No, I carried my requisition signed and ready for business. Also I took along a French sergeant to aid at the French factories. As a result, I obtained several boxes of parts which I had shipped and sent back as baggage."[39] He would make a similar trip to Paris in September.

The War Winds Down

As the war moved into the late summer and early fall, the German army, having failed to achieve its objectives along the Western Front during three major offensives in the summer of 1918, fell back close to the Hindenburg Line in September. Assisted by the growing number of American forces, the British and French armies recaptured most of the ground lost to the Germans during the summer. Clayton followed these Allied gains with interest, particularly the success of the first all-American offensive of the war on September 12 at St. Mihiel, south of Verdun. The Americans took the Germans by surprise and captured 13,251 prisoners and 466 guns.[40] "We in the rear all knew it—the Americans, so I'm informed, told the Bosch we were coming there—and yet it was one of the most perfect feats of the war against one of the formidable German defenses," he wrote on October 13. "The whole affair happened in just 19 hours."[41]

These advances gave Clayton hope that the war was coming to an end. "The bottom has dropped out of Germany in the last month and she seems to know it," he wrote. Then, with some bitterness, he criticized those who would make a quick peace with Germany before the German army was completely driven out of France and Belgium. "And it may be the sorrow of many a murdered soldier's brother that right now when they are giving the Germans hell he [Germany] is allowed to escape his penalty. But for all in all it's probably—carry on, carry on, for fate is not in my hands nor is it my privilege to discuss such matters. . . . The poor French people are raving— whooping and yelling and yet some have that self-satisfied smile—'I told you so all the time'—while any ordinary well behaved man could not figure how a nation which defended itself by the point of its sword and by the help of God would sign a treaty or plea for an Armistice before she [Germany] was forced off of captured soil."[42]

With the exception of his letters home describing his time in Flanders, Clayton rarely spoke about the horrors of war, preferring to talk in the abstract about battles and troop movements. This may have been due to the requirements of censorship or to an effort to shield his parents from the realities of what happened to men on the battlefield. However, Clayton did share one bloody story with his father in a letter written after his September parts-hunting trip to Paris.

During that trip, Clayton linked up with several old friends from Texas A&M, including Max D. Gilfillan, a former college athlete who was serving as a second lieutenant in the marines. Clayton learned that Gilfillan had been badly wounded at the Battle of Château-Thierry. "He was shot in the hip by a machine gun bullet which exploded a clip of cartridges on his hip and took out an awful chunk of muscle and flesh—cutting an artery, which left him unconscious in a few minutes from loss of blood. Luckily a friend . . . found him, knew where to tie the bandage, did it [and] carried him back to shelter. But the friend was himself killed within seconds."[43] Gilfillan told Clayton that another schoolmate, Direck Milner, was wounded so badly that he was no longer on active duty and that Thomas R. Brailsford, a former baseball player at A&M, was missing in action. Brailsford was the first Marine Corps officer to be declared missing in the war.[44]

Going Home

With the signing of the armistice on November 11, Clayton's duties at Clermont-Ferrand quickly changed from instruction to tying up loose ends.[45] Because of his ability to "make do," he was one of five artillery officers ordered to remain at the center to "put the final touches on clearing out the Center. . . . To be frank with you, I am the only artillery officer here who has a line on the red tape . . . so as a result I have had to make out [and] arrange details and prepare all things for the transfer of all . . . property."[46]

Having learned during this period that his promotion to first lieutenant had once more been denied, Clayton again expressed his frustration to his father. "Well, this is not unusual for me—it's my ordinary run of luck and I am now getting accustomed to such [a] run of affairs and simply call it fate. . . . Now the war is finished. I have done my share—among [the] 1st to volunteer—1st to come over—I am now asking for immediate separation from the Army with a reappointment in the reserve. I expect to be sailing home within 6 weeks." After the war, it was with a touch of rancor that Clayton declined an appointment as a second lieutenant in the Coast Artillery Section Officers' Reserve. He replied to the army's offer, "If merit receives recognition in the Army, I refer you to my record and to my qualification card, both of which are to my best knowledge very good or excellent. My experience as an officer, dating back from Aug. the 15th, 1917 with my early volunteer departure for France—Sept. 12th, 1917 and late return—Feb. 21st, 1919, certainly deserves more recognition than the experience of those gentlemen who just completed the last Officer's Training Camps."[47]

As the days dragged on at Clermont-Ferrand, Clayton's thoughts turned to his future after the war. He wrote his father on December 12 that there were some business opportunities in France, but that he could not act on them because he would have to return to the United States before being released from the army. "Also, the chances over here required either capital or the faculty of speaking excellent French—neither of these I have." With his typical confidence in American superiority, Clayton believed that the French were eager to hire Americans because of their business skills (Americans had "the best business methods in the world") but confessed that his only business experience was what he had learned in the army, particularly at Clermont-Ferrand."[48]

Clayton told his father that he had been talking with a Lieutenant Holt, the son of the founder of the Holt Manufacturing Company that made the Caterpillar tractors used by the army, about going into that company's sales department. Benjamin Holt and his brother Charles had incorporated Holt Manufacturing in 1892 in Stockton, California. The company manufactured first horse-drawn and later steam-engine combines. In 1904, the Holt company successfully tested the first track-type "Caterpillar" crawler tractor, which was needed to farm the soft, spongy soil of the Sacramento-San Joaquin River Delta between Stockton and San Francisco.[49]

"Well, he's a nice chap but I am not sure that his word would procure me a place, for to his father he may still seem as just a boy. You know what I mean—his father probably does his own picking and choosing of salesmen." In this long letter about his future, Clayton described for his father his duties and level of responsibility at Organization and Training Center No. 3, apparently to demonstrate his competence and skill. "I am now 2nd in charge of auto instruction—which is somewhat similar to a cashier and assistant cashier in a bank—as you know the latter does most of the work." Clayton expressed the hope that O. W. could use the information to "line me up with some opportunity." He closed the letter by indicating some interest in returning to A&M to pursue more education in either mechanical or electrical engineering, but was unsure if that was wise because he felt that those professions were overcrowded and wages were poor. "The business end of the deal is the place to make money," he concluded.[50]

As Clayton struggled with deciding his future course, his father offered his own advice. In a letter to Clayton's brother J. C., O. W. shared his opinion that Clayton's army training, coupled with his electrical engineering background, would seem to position him perfectly for a job in the automobile business in the United States. "So I have suggested to him that he hold on in Europe for a time if he can do so by volunteering to remain longer [and] keep his salary [and] experience going—until we get ready at home to give him a good chance for employment." However, if Clayton were ordered home, O. W. felt he should return to Texas A&M, finish his degree, and take courses toward a second degree in mechanical engineering, apparently because he had expressed an interest in becoming an inventor. "With these qualifications he could secure employment at good wages—even if he never made good as an inventor—and without them he would be crippled in trying to invent mechanical devices."[51]

In a long letter to his father written on December 22, Clayton continued to speculate about his future after the war. He was now convinced that immediate separation from the army was in his best interest because "many industries have been waiting for the War to finish to start up again—and . . . the early bird gets the worm." He told his father that his port of debarkation upon arriving in the United States would be New York (it was actually Philadelphia), and that he believed that New York would be the perfect place to contact potential employers. "It is better," Clayton wrote, "to wait on things to develop with your name on a waiting list than to arrive too late."

Acknowledging the suggestion by O. W. that Clayton return to A&M to finish his degree, Clayton said he would do just that "if things do not develop as rapidly as I think." Clayton also revisited the idea of working in France. Agreeing with his father's suggestion, he now believed that through more study he could become competent in speaking French, and that this skill, coupled with his experience in electric engineering and instruction in automobiles, would make him an ideal representative for American automobile manufacturers doing business in France after the war.[52]

Reinforcing his newfound enthusiasm for working in France was Clayton's belief that the country and other international markets would be a profitable future source of business for American industry. He argued that American industrial capacity had been ramped up by war and was ready to turn to producing peacetime goods, that the American merchant marine was in place to transport them, and that there would be little competition from other countries. He cited the Liberty motor, which had been developed a year earlier in the United States to power airplanes, as an example of an American product that could be sold in France and other countries.[53]

The development and manufacture of the Liberty airplane engine was one of America's finest industrial production projects of the war. The engine had standardized, interchangeable, mass-produced parts, making it easier to repair in the field than European engines. More than 20,000 Liberty engines were produced by the time of the armistice. The engine was manufactured by automobile companies but was never adapted for automobiles. However, at least one commentator acknowledged that the rapid development of aircraft engines during the war led to "new ideas" in automotive engine material, design, and production.[54]

Clayton pointed out to his father that most of the contracts for reconstruction of northern France had been let to American and English contractors. He made an elaborate argument in which he estimated that some $50 million was spent monthly in French stores, cafés, and other enterprises by French, American, and British soldiers and civilians. Clayton concluded that the French business class had plenty of money: "Should anyone doubt that the café owners and merchants of France are not doing business and making money, just let them shop in Paris or any other city or village where American troops are and just single out any one place—keep his eyes open for the amount of trade done and at what prices."[55]

But as he entered the New Year, Clayton abandoned the hope of a business career in France. In his letter of January 4, 1919, he told his father that he had decided to study law at Columbia University. "My desire in taking this step is to come in with you where I have a chance for a start—my only chance so far—and I'm getting along where my steps must be sure now." And, apparently to reassure O. W. that he was a changed man, Clayton wrote, "I'm not old yet but the time for fooling around has long been past. I have certain habits, which I can and *will* break. These habits are detrimental to any man of good standing. I know it! But I have called 'quits' on them. My eyes are the only dubious thing in my road and I believe smoking affects them."[56]

Clayton received three orders between January 11 and January 13, 1919, relieving him of duty at Clermont-Ferrand and ordering him to proceed first to Angers and then to Blois, located midway between Orleans and Tours. He reported on January 17 to Blois, where his return to the United States was processed, but it was not until the first week of February that Clayton gradually began to move up on the sailing lists for embarkation home. The long wait was frustrating. He remarked in a letter that his friend Lieutenant Holt, who had left Clermont-Ferrand ten days after Clayton, was by February 7 already eight days at sea.[57]

Clayton was in a thoughtful mood as he left the harbor of Brest, France, on that winter afternoon aboard the *Northland*. "It is always sad, this parting," he later wrote home. "One officer next to me remarked, 'This is my last sight of France—sunny and muddy France!'" Wondering if he would ever return "when the effect of war was not evident," Clayton examined his attitudes toward the country and its people:

> All of my letters have seemed to show some prejudice against
> the habits and customs of the French. It may be a matter of
> circumstance and probably if I had not the acquaintance of an
> excellent French family in Paris, I might be even bitter against
> the people. But as a whole I've tried to make my observances
> of a people new to me as fair as an interested inexperienced
> American could. Naturally an American thinks the Ameri-
> can brand better than anything else so he cannot help being
> prejudiced. However, the French were very patient with us
> in our prejudices and often agreed with us. I end this tangle
> of thoughts by merely considering [that] every people has its
> goods and bads and I could at present follow only my preju-
> dices with a distant hope of again seeing France when the rose
> is blooming and not drenched with blood.[58]

The crossing was not without incident as the *Northland* lost one of its boilers, encountered several days of high seas and rough weather, and suffered two deaths aboard, one from influenza and the other from meningitis. The trip was particularly hard on the enlisted men, who were housed in stuffy, "odoriferous" conditions on the lower decks. Finally, the *Northland* arrived at the mouth of the Delaware River on the morning of February 21, where she was greeted by a boat carrying five newspaper reporters. As she sailed up the river, factory whistles and boat whistles sounded to honor the men onboard, and a launch with a welcoming committee of "good looking girls" and a band escorted the ship into its berth at the Philadelphia pier.

Clayton was moved by the fanfare. "I realized this was all for us—and they were really glad to see us," he wrote. "They looked at the boys who stood there and who had so nearly lost their lives time and again, some even then with gassed and ruined lungs, others with fingers, arms and legs gone—and [as I thought about] those who had made the supreme sacrifice and would never receive a 'welcome home' except probably by the Father above—a lump came up in my throat, the tears came out of my eyes, and it was hard to keep a straight face."[59]

CHAPTER 6

Searching for Oil West of the Pecos

C layton came back in 1919 to a country that was jubilant over the
return of its men from war but uncertain what to do with them.
Although greeted with bands, parades, and the promise of bonuses
for their time in service, the doughboys found few opportunities for employment. Just as the country had been ill prepared for war, it was ill prepared
for peace. With the sudden cessation of wartime contracts, many industries
had a difficult time converting to peacetime production, leading to massive
layoffs and a declining labor market for the returning veterans.[1]

The country was in recession, and as Clayton would quickly learn, jobs
were hard to come by. As planned, he began his job search in New York
City, and on February 26 Clayton and his friend Lieutenant Holt met with a
representative of the Holt Company. According to a February 27 letter to his
father, Clayton's intention in having the meeting was not to get a job with
the company, which manufactured the Caterpillar tractor, but to determine
if agricultural land along the Atlantic coast and in the southeastern United
States would one day be a good investment.

"Now don't think that I'm speculating like some damn fool pretending
to have money," Clayton told his father. "This is all good information which
may help some day. I [will] find out just about how tractors are going—[and]
make [an] acquaintance in an interesting and business way who may be of
use to me some day. And increase my knowledge of our country." Clayton
told his father that the Holt representative believed that Texas and Mexico
"are going to be big fields for the tractor. So do I and I've got my eye on some
parts of the Brazos bottom land that hasn't been cleared yet."[2]

Having abandoned the idea of studying law at Columbia, Clayton began his serious job-hunting a few days later. He looked up a business acquaintance of his father's who encouraged him to send letters of application
to a number of companies, including International Banking Corporation,
American International Corporation, Guggenheim Exploration, General

Clayton and sister Kathryn after he returned home from France in 1919.

Electric, National City Bank, and Westinghouse. Only American International requested a personal interview. The corporation owned a number of companies that sold American products in Europe, South America, and Asia and had gross earnings of $12 million in 1919 (equivalent to $149 million in 2009 using the Consumer Price Index).[3]

Unfortunately, the man with whom Clayton expected to interview was out of the office because of the death of a child. "So I spoke to his assistant, who informed me that his boss must have had something special in his mind. The only opening he himself knew of which might 'pertain' to me was [for a] young engineer to work under the wing of their chief engineer for special training here in N.Y. City with the ultimate result of going abroad and handling something big! One of the requisites was a foreign language.

This he said started with a '*measly*' salary of $125 with a fine chance for advancement.... But as he did not know of his boss's intentions in regard to me [he said to] call again if possible and *see the boss*. This I shall do tomorrow."[4] Clayton was never able to meet his intended interviewer at American International, however, and his other application letters went unanswered. "From my inquiries I find I'm too late arriving in U.S. as all big jobs on exportation have been filled," he wrote his father. "A slow answer [to the employment inquiries] is a good indication."

Like any young man returning from the war, Clayton did not spend all his time in New York solely on a job search. A friend of his father's treated him to a night at the opera, where he saw Enrico Caruso sing the role of Rodolfo in *La Boheme*. Caruso was one of the most popular singers in America. A few months before Clayton returned from Europe, the forty-five-year-old singer had married Dorothy Park Benjamin of New York, who was twenty years his junior. "Such singing—it is hardly possible to conceive of a voice as perfect as Caruso. Although he has been singing for over 25 years, he still plays and cuts up on the stage like a big boy.... Being the 1st time I ever saw Caruso I didn't know whether it was his habit to cut up or his new bride affected him thusly."[5] After the opera, Clayton's host took him to a cabaret and treated him to dinner and cigars. He attended other shows and musicals as well, and his time in New York lightened his wallet significantly. Between going to shows and dining extravagantly, "which I didn't have a chance to do during the past 19 months," he wrote his father, "I have spent $150 which I can ill afford to be deprived of. This rather pushes my steps homeward. I plan on leaving Tuesday."[6]

Clayton left New York on March 11, making a brief stop in Washington, DC, to pick up a sixty-dollar government bonus for his time in military service. He also inquired whether his military service would reduce his waiting time for a homestead. The bonus money had been authorized by Congress in the Revenue Act of 1919, passed just a few days after Clayton returned to the United States. The amount was based on two months of salary for an enlisted man. It was to be used by soldiers to bridge their transition from the military to civilian life. Congress had also extended to veterans the right, under the Homestead Act, to count military service in lieu of residence toward a homestead exemption for tax purposes. A number of other proposals were discussed in Congress, by officials in the federal government and in some states, to provide returning veterans with land for homesteading. With a few minor exceptions, primarily in South Dakota, most of these schemes did not get off the ground.[7]

"After Washington I may come straight home," Clayton wrote his father. "I still have a strong leaning toward law and I think it my best future chance.

Rather than waste any more money on schools, couldn't I study up and take the bar exam. Would it be asking too much of you to give me a little coaching on such? When I get home we'll go over this together.... I have my fill of shows and such and am ready to come back to Stockton but I never liked that March and April breeze of west Texas which blows boulders from mountain to mountain. Although this spring it may be better as Europe and America have had rather extraordinary mild weather this winter."[8]

From Washington, Clayton traveled by train to San Antonio. After spending a few days there, he caught a train to Alpine and then over to Fort Stockton, arriving on or about March 22. "I thought they would have someone to meet me, but no one met me," Clayton recalled later. "That was kind of a disappointment. The train . . . from Alpine came early in the morning—even my parents weren't up. I was really glad to be home."[9]

First Smell of Oil

Throughout the spring and summer of 1919, Clayton spent his days working at different odd jobs, including drawing maps, repairing automobiles, and wiring houses.[10] But by October, he had decided to follow the scent of oil to north Central Texas, where two years earlier the Texas and Pacific Coal Company had discovered significant oil deposits southwest of Ranger, located roughly midway between Fort Worth and Abilene.

Since the turn of the century, oil fever had slowly spread through Texas. The first important Texas oil discovery was at the famous Spindletop site near Beaumont, where in 1901 drillers hit what was at that time the greatest gusher ever found. As production was being developed along the Gulf Coast, oil was discovered in 1902 in North Texas east of Wichita Falls, near the border with Oklahoma, and by 1905 production there totaled 500 barrels a day. In 1907, oil and natural gas production was being developed at Petrolia, nineteen miles northwest of Wichita Falls, and there was other significant production at Electra, about thirty miles west of Wichita Falls. The new finds stimulated drilling throughout North Texas, and by the end of 1919, North Texas oil output for the year was more than 53 million barrels, with a market value of $120 million.[11]

This discovery of oil at Ranger was followed by the development of fields to the south at Desdemona (originally named Hogtown), thirty miles to the north near Caddo, and west of Caddo at Breckenridge. The Breckenridge production increased from a little more than 36,000 barrels in 1917 to a high of more than 31 million barrels in 1921, and the town's population also increased dramatically as speculators and others, both scrupulous and unscrupulous, flocked into the area seeking new opportunities or simply a fresh start.[12] News of these developments filtered south to Fort Stockton, and

Clayton was lured to north Central Texas in the fall of 1919 by the promise of steady employment. He was hired as an assistant electrical engineer by the newly created Oil Belt Power Company in Eastland that supplied electricity to the towns of Ranger, Breckenridge, Eastland, Gorman, and DeLeon. His first job with the company was to map out a route for reading the electric meters.[13]

Despite his son's having landed a full-time job, O. W. was not happy. "I am not well pleased with his choice of $125 as Electrical Engineer, as he will save nothing at the latter job & might have saved $8 per day at the former [it is unclear from the letter what the former job was]. In one sense however he is doing well, as he is getting in line with a profession, although he will be several years working up to a decent salary. One must have a business or profession in life."[14]

As new wells were dug in North Texas, adjacent land rose in value; speculators secured oil leases in the hope that they, too, could tap into the black gold flowing underground. Clayton was among those who succumbed to oil speculation fever and that fall paid $750, money he had saved during the war, to buy oil leases on five acres of land north of Breckenridge. O. W. once again expressed his dissatisfaction with his son. "I am not very well pleased, as I fear he had not been there long enough to understand the proper value of his purchase," O. W. wrote J. C. "He bought from a former A&M friend, [and] it may be that he was not as badly worked as if some other party had taken him in. But it is a poor plan to play another man's game. Learn your game before playing any stakes on it." Clayton later said that he lost his investment in the leases.[15]

There is not much information in the family record about Clayton's time in Eastland. There are no letters from Clayton during that period, perhaps because he was busy at his new job and did not have time to write. "He is almost as poor a correspondent as you are," O. W. complained to J. C. Clayton apparently did find time for some fun outside of work, however, once telling O. W. that he had participated in an Elks Club show in Eastland, singing songs and acting in a play. "Too much society I am afraid," groused O. W.[16]

While Clayton was laboring in the oil fields near Eastland, speculators were becoming increasingly interested in the prospect of finding oil near Clayton's hometown of Fort Stockton.[17] For several years local residents had been aware of oil on the surface at a ranch about fourteen miles north of town, but drilling in the area in 1900 had been unsuccessful. Several companies drilled wells around Fort Stockton in 1916 and 1917, but again without success. Prospects brightened slightly in 1918, when E. A. Reilly and C. E. Menzie, experienced operators from Kansas and Oklahoma, independently drilled for oil near the town of Sheffield on Pecos County's eastern border.

The Williams family on the porch of O. W.'s house along officers' row in Fort Stockton. The photo was taken while Clayton was working as a highway engineer near Sanderson, Texas, from 1922–24. Clayton is seated in front next to his sister Kathryn. The others, from left, are Ermine, O. W., Sallie, and Waldo.

Their activities were followed intensely by the residents of Fort Stockton, but neither man struck oil, and by June of 1921, both had pulled out.

However, geologists who visited the area were convinced that oil deposits lay beneath the surface. A professional paper published by geologist H. E. Peterson of Berkeley, California, stated that surface configurations were favorable for significant oil deposits in an area along the Pecos River between the towns of Sheffield and Girvin. Peterson leased a large number of acres from landowners in the area and contracted with several individuals in 1918 and 1919 to drill wells on some of his leases. Unfortunately, Peterson overextended himself, failed to start a single well, and had to surrender the leases. Despite his lack of success, his activity stimulated rumors that oil would soon be found in Pecos County, and Fort Stockton became the gathering place for operators, speculators, and others in the oil business. In 1920, Clayton quit his job in Eastland to return home to "make my fortune."[18]

Wonder and Miracle Wells

Clayton was drawn home after hearing about two producing wells in the Turney Ranch area north of Fort Stockton, drilled by J. W. Grant of Pittsburgh, vice president and general manager of the Grant Oil Corporation. The first well was named the Wonder well and the second, the

Miracle well. The Miracle well produced an estimated 5,000 barrels of oil on the day of discovery. "The telegraph lines were busy," Clayton later recalled. "People all over the country were advised of this Miracle well. Newspaper men bundled up their bag and typewriter and headed for Fort Stockton along with all the loose speculators in the region." However, the output from both the Wonder and Miracle wells quickly declined and within a week suffered from intrusions of water. Efforts to improve oil production failed and the wells were abandoned.[19]

The fact that the Miracle well had produced oil, however briefly, was enough to stimulate more drilling throughout Pecos County, and by June some eighteen companies were actively drilling. However, according to Permian Basin oil historian Samuel D. Myres, none were major oil companies. "To the few geologists of the large companies who surveyed the territory at the time, the chances of finding oil in the vast semi-arid wilderness appeared exceedingly remote. The majors were therefore willing to leave the entire area to the adventurous wildcatter and small-company operator."[20] None of these small operators succeeded, and by January 1923, the Pecos County oil boom had gone bust.

Other events during this period contributed to a feeling among oilmen that the area west of the Pecos River would never produce commercial oil. One involved the Troy well, which was begun in 1921 about one mile east of the Miracle well. According to Myres, the operation was owned by investors from Wyoming; W. B. Troy was the driller and field manager. The well was drilled to a depth of 1,125 feet before the company ran out of money. Believing that the oil would be located at a shallow depth, as it had at the Miracle well, the investors refused to finance deeper drilling. Fired, then rehired when the investors found no one to take his place, Troy located a new investor who financed further drilling. When the well reached 3,000 feet it was reported to have come in, but it was never completed, leading to a persistent rumor that it had been "salted" with oil brought over from J. W. Grant's Miracle well.[21] Clayton recounted part of the story years later. He said that he had inspected the well personally and found that Troy "had successfully drilled down to a depth between two and three thousand feet when the report got out that [the well] had struck oil and flowed over the derrick." Clayton found that "the derrick was covered with oil but it was also boarded up so that it took considerable effort to get inside and take a look at the well casing. The top of the casing was fitted with a swage nipple and gate.[22] The gate was locked, but a small amount of sulfur gas was escaping from the joint between the nipple and the casing."

Years later, Clayton would learn that the test well had struck a strong flow of sulfur water, which in those days meant that there was no chance of striking oil. "Under those circumstances," he wrote, "Troy and Grant [had] pumped a large amount of oil on top of the sulfur water. By swaging and reducing the casing outlet, this oil was spouted out through the derrick from the pressure of the flowing sulfur water under it. Troy's financial supporters came down to view their prospects. The cat was out of the bag and they found Troy gone to parts unknown."[23]

Another blow to Pecos County's reputation as a source of oil was an article that appeared in the *Oil and Gas Journal* in 1919. Written by geologist Charles C. Coulter, the article was titled "Petroleum 'Graveyard Area' of Texas." Based on a three-month survey, Coulter predicted that no oil would be commercially produced in a 30,000-square-mile rectangle extending from the Panhandle of Texas to an area northeast of Fort Stockton and the Pecos River between Upton and Menard Counties. "That picture he made looked dark and dismal for those who were hopeful for the oil prospects of Pecos County," said Clayton years later.[24]

Clayton tried a number of ways to make his fortune in Fort Stockton, including trading in oil leases with his brother J. C., drawing maps (including the first city map of Fort Stockton), leasing and running a silent picture show ("accompanied by the joyous strains of a player piano"), and being elected Pecos County surveyor.[25] None of these jobs produced sufficient income, however, and in 1921 Clayton accepted a position as the "instrument man" for a surveying crew working a highway-paving project west from Anthony in southeast New Mexico.[26]

It was while working on this project that Clayton had what he regarded as one of the few serious fights of his life. He had been examining concrete to see if it had the correct amount of fill and discovered that the sand being used had too much dirt in it. So he turned it down. "The fellow running the outfit [contracting crew] told me he was going to put me and my helper in the hospital if we didn't watch out. I jumped him about it. I hit at him and hit over his shoulder and I closed in on his neck, but then they separated us. I invited him across the line but he wouldn't come. He had more sense."[27]

After about a year of work in New Mexico, with his father's assistance Clayton landed a job in Sanderson, Texas, surveying a state highway through Terrell County for $200 a month.[28] Sanderson is located about sixty-five miles southeast of Fort Stockton. His contract called for six months of work beginning December 15, 1922. "His party have a camp in Sanderson and go out to work and return in cars," O. W. wrote J. C. "He pays $35 a month for board and uses a tent for his room. Says he is pleased with his job and his chief."[29] While at Sanderson, Clayton worked in his spare time on the first real waterworks system for Fort Stockton. He was resident engineer on the early stages of its construction.[30]

Working with Waldo

Clayton worked on the highway project out of Sanderson for nearly two years, but he was also keen to explore other means of making money. Much to O. W.'s chagrin, he decided in the spring of 1924 to invest with his older brother Waldo in a plan to farm cotton on irrigated land near the town of Piedras Negras in Mexico. Waldo was already sharecropping cotton on 300 acres in the State of Coahuila, Mexico, near the town of El Moral, on the Rio Grande about fourteen miles northwest of Piedras Negras. He now asked Clayton to join him in clearing and farming an additional 500 acres in the same area.

"I am not well satisfied with the idea," O. W. wrote to J. C. "First the pink Boll worm is rampant in that Territory & and all Texas quarantines against cotton & cotton seed from that territory. That means they could only sell their stuff in Mexico. Second. Everybody & his brother farms cotton this year. That means a big crop this summer & a great drop in the price next fall. Third. It also means that every Mexican laborer is already grabbed by somebody & Waldo comes in too late to get men to work his lands. So I am afraid the scheme is bad."[31]

O. W.'s concern was well-founded. Having graduated from Carthage College in 1904 with an engineering degree, Waldo had held a number of jobs before trying his hand at cotton farming but had never stayed with anything very long. He married Olive E. Strickler in 1913, and the couple settled in Fort Stockton, where Waldo worked on numerous irrigation projects and was briefly the town's postmaster. The couple left Fort Stockton seven years later when Waldo became manager of the San Ysidro Ranch in Coahuila.[32]

"He once said to me, anything he liked doing he never could make a living at," wrote his daughter Mari Helen Williams Schultz.[33] A colorful and adventurous personality, Waldo was a great storyteller who claimed many adventures. One story he told involved the controversial 1908 London Olympic 400-meter dash, which the American runner J. C. Carpenter won; but he was disqualified for jostling a British runner with his elbow. Waldo claimed that he had run in the race under Carpenter's name. He explained the deception by saying he had once been paid to play baseball, which made him a professional athlete and unable to compete in the games as an amateur under his own name.

Waldo also told a story about serving as a machine gunner fighting against the Mexican government in the army of the revolutionary Pancho Villa. Clayton's nephew, David Walker, says that Waldo told him he had joined the revolutionaries after being spurned by a woman. According to

Walker, Waldo's love interest, ironically, was one of the daughters of A. J. Royal, the Pecos County sheriff and sworn enemy of O. W., who had been murdered in 1894 in the Fort Stockton courthouse.[34] Waldo said that in one battle with the revolutionaries, he had been manning a machine gun when his comrades pulled back, leaving him and his assistant gunner exposed to fire, with bullets ricocheting wildly off the front and back of the deflector shield. The assistant was killed and Waldo ran down an arroyo and eventually crossed the Rio Grande back into Texas.

Clayton relayed these and other stories about Waldo's adventures to O. W. He said that Waldo had had a brief career before World War I as a pilot in a "flying circus." In flying circuses aviators, also called "barnstormers," performed stunts with airplanes, such as flying upside down and walking on the wings while the plane was in the air. Such aerobatics were limited before World War I, but the return of military aviators after the war and the availability of used aircraft saw barnstorming reach its peak in the 1920s.[35] Clayton said Waldo avoided an airplane crash when his boss, who died in the crash, grounded him before making the fatal flight. Clayton also told O. W. that Waldo had raced against the Blackfoot Indians in Montana and killed a foreman while working on a track-laying project for a railroad. O. W. seemed somewhat skeptical about the veracity of some of these stories, but said, "Our family has so long been among the Adventurers on the far frontiers that hardly any foolish stunt should surprise me."[36]

"Waldo was a runner," said Clayton's son Claytie. "He ran to Alpine one time [the distance from Fort Stockton to Alpine is more than sixty miles]. And he was artistic. His drawing was probably much superior to my dad's."[37]

It is difficult to determine whether Waldo's stories are all truth, partial truth, or complete fiction. Concerning Waldo's claim to have run in the Olympics, the *New York Times* reported that J. C. Carpenter was from Cornell and had been a high school runner in Washington, DC. As Carpenter was a competitive runner and probably well known on sight to his fellow competitors and race officials, it seems highly unlikely that Waldo could have substituted himself as Carpenter. In "confessing" the subterfuge, Waldo said years after the race that someone had recognized him from a photo taken at the finish, causing him to publicly reveal the "truth." O. W. told another variation of the Olympics story and said that Waldo ran races in Montana and New York for money "to cover his time and expenses" and believed that this would cause him to be disqualified from participating in the Olympics. O. W. also said that Mel Shepherd and Ted Meredith had run in the 1908 games and previously competed against Waldo, so both men would have recognized him. However, only Shepherd ran in 1908; Meredith did not compete in the games until 1912.[38]

The story about the flying circus is also suspect. In retelling it, O. W. said that one of the members of the circus was a famous French aviator named "Fonk" who was an ace in the First World War. This was no doubt Réne Paul Fonck, who had seventy-five confirmed kills during the war, making him the highest-scoring ace for France and the Allies. However, Fonck did not begin flight training until 1915, so it is improbable that he was flying for a circus prior to the war.[39]

Tall tales aside, Waldo was intimately familiar with the land bordering both sides of the Rio Grande. While a teenager, he, like his brother, had accompanied his father on surveying trips into the Big Bend area of Texas and had also worked on the construction of a portion of the KCM & O (Kansas City, Mexico, and Orient) railway in Mexico. By 1924, Waldo and his family, which included three children, lived in Eagle Pass on the Texas side of the border, and he worked at various engineering projects in the area in addition to farming cotton.[40]

O. W. was deeply uneasy about Waldo's ability to support himself and his family. "I have been much concerned about his affairs for some time, as his family seems to cost him altogether too much, especially since his income has been practically cut down to nothing since the Mexican Revolution began," O. W. wrote J. C. "It costs him $200 a month for living expenses, and lately he has earned $100 a month or less, as he told Clayton. Now the time for a poor man with a family to save money is while the family is small, because when the children get up to 10 to 18, the education for them costs heavily. Yet Waldo is on the very verge of that period right now with nothing but debt on hand, & more debt threatening."[41]

In May 1924, Suzanne Sykes-Gaubert, Clayton's "godmother" in Paris, decided to help finance the cotton-growing venture. Clayton asked O. W. to prepare a power of attorney authorizing Clayton to handle her money and business affairs in Texas. "So it would seem that he and Waldo are going in the cotton scheme over the river from Eagle Pass and will use her money," O. W. wrote J. C. "I have not favored this plan, but it seems that I learned of it too late to check it."[42] In early June Suzanne sent word to the Williams family that she was preparing to leave Paris for Texas, and by mid-July she was staying with Waldo and his family in Eagle Pass. The plan called for Clayton and Waldo to manage the cotton-farming venture and for Suzanne to finance it with $7,000. O. W. expected he would be asked to draw up the final contract, and he continued to fret.

"I wish that Clayton would realize that it is not so much what one makes as what is saved that counts in the race for life earnings," he wrote J. C.

Suzanne Sykes-Gaubert, the daughter of the French family that befriended Clayton in Paris, shown with Waldo's wife Olive and their children (from left) Olivia, Oscar Waldo III, and Mary Helen. Suzanne came to visit the Williams family after the war and invested in an unsuccessful cotton-farming venture with Waldo and Clayton.

"Here he is nearly 30 years old & without any money or property ahead. And Waldo is in even worse shape, for he has a family of promising young-sters and nothing but his daily labor to keep these chaps through childhood and education. However I went through the same experience and survived, which may account for the fact that I felt the dangers at the time and later, and so I am more concerned about their future than they are."[43]

As the financial arrangements took shape, there was some disagreement among the parties as to how much each should invest. Suzanne Sykes-Gaubert agreed to put money into the project but wanted Clayton and Waldo to invest more than simply their time and Waldo's labor. O. W. was now hopeful that the agreement would fall through.

"Her mother had encouraged her, but her father had opposed the propo-sition, and so far not a dollar of that $7000 has been paid to the Bank here," wrote O. W. following a visit to Fort Stockton by Clayton, Waldo and Su-zanne. "It is probable that the father is holding back the transfer of the money, and you may have heard that the French Government has put such obstacles in the way of sending money out of France that Suzanne, in order to get a permit to send this money out, had to represent that she owed the $7000 to Clayton. I suppose that the matter will be finally thrashed out at

Eagle Pass, where Madame Suzanne has gone to stay with Waldo until matters are decided, after which she will return to Paris."[44]

After borrowing $500 from Clayton for her travel expenses, Suzanne sailed for Paris on September 14, and eight days later O. W. was notified by his bank that $7,000 had been telegraphed from the Chase National Bank of New York to Clayton's credit. "I had supposed that she had abandoned the idea of the cotton farming plan, but it seems not," wrote O. W. "Clayton came in last night & tells me that Waldo must soon begin to prepare the 500 acres of land for farming—much of it being in the bush—& requiring not merely the ditching & leveling, but also the grubbing." The land would be irrigated by pumping water from the Rio Grande. "I am very much disposed to doubt the safety of that pump," wrote O. W. with his usual pessimism. "It never has done much heretofore, & if it breaks down when needed, the crop is lost."[45]

The formal agreement for the cotton venture was signed by Clayton and Waldo in Texas on November 1, 1924, and by Suzanne in Paris on November 18. It called for Waldo to manage the business and receive $200 a month for his expenses. Waldo was further required to invest the equivalent of $6,000 either in equipment, stock, cash, or a combination of these. Clayton was required to invest $6,000 in cash as his share, and Suzanne would also invest $6,000. However, disaster struck the project later that year, and it wasn't the pump that failed Waldo. The State of Coahuila issued an order to confiscate several thousand acres of land near Moral, including the 500 acres leased by Waldo and Clayton. The land would be redistributed to small farmers. The movement in Mexico to redistribute agricultural land from the large haciendas back to the villages had been an important driving force of the revolutionary period of 1910–20, and land reform legislation was incorporated into the 1917 Mexican Constitution.[46]

Waldo had invested money in mules, harnesses, and plows, but no farming could be done on the land near Moral. He appealed to the American Consul in Eagle Pass to persuade the Mexican government to reimburse him for his losses, but O. W. doubted that any relief would be forthcoming. "In Mexico, practically bankrupt from repeated insurrections, and with a long-seated grudge against 'gringos,' it is almost hopeless. The money spent, and the time and trouble lost, would mount up in a few years to more than the claim. So I think Waldo had best pocket the loss he has encountered to himself & to Madame Suzanne, & do what cotton farming he needs on this side of the Jordan [Rio Grande]"[47]

Undaunted, Waldo made arrangements to lease several hundred acres of land near El Remolino, Mexico, about thirty-four miles west of Piedras Negras, but this venture also failed when the crop was devastated by the cotton

Waldo and Olive on the land near Piedras Negras in Mexico where Waldo, Clayton, and Suzanne Sykes-Gaubert invested in the unsuccessful cotton project.

fleahopper, one of many insect pests that prey on cotton.[48] "I think I should have educated my boys for the oil business, as it seems a surer source of income than anything Waldo & Clayton have yet found," concluded O. W.[49] According to David Walker, Clayton felt so bad about Suzanne's lost investment that he personally paid her back out of his own pocket.[50]

Santa Rita No. 1

B ut Clayton's luck was not all bad. His work at Sanderson placed him in contact with Texon Oil and Land Company employees who were working their way southwest from Big Lake, Texas, about 100 miles to the north, in the search for oil. In May of 1923, Texon brought in the now-famous Santa Rita No. 1 well located on land owned by the University of Texas in Reagan County about four and one-half miles west of Best. The well was named for Santa Rita after Frank Pickrell, who was executive vice-president of Texon, visited a group of Catholic women in New York who had invested in Group No. 1 stock before the well was brought in. Somewhat nervous about the risky nature of their investment, they consulted their priest, who suggested they put their faith in Santa Rita, the patron saint of the impossible. The women gave Pickrell a red rose that had been blessed by the priest in the name of the saint. They asked Pickrell to sprinkle the rose petals over the derrick and christen the well Santa Rita.[51]

Santa Rita No. 1 was the first of a series of wells that unlocked the oil and natural gas resources of the Big Lake Oil Field, firmly confirming the promise of the Permian Basin area as a major petroleum resource. The basin takes its name from the ancient sea deposits formed some 250 to 300 million years ago during the Permian geologic period. The basin occupies an area of West Texas and southeastern New Mexico that covers approximately 100,000 square miles of land.[52]

"Santa Rita No. 1 not only extended the potential oil province of West Texas some ninety miles to the southwest of Mitchell County, but it also opened one of the great oil fields of the state," wrote Myres. "The field made fortunes for scores of investors, saved the Kansas City, Mexico and Orient Railroad from almost certain extinction, and laid the foundations of the great wealth of the University of Texas, a struggling institution housed at the time mostly in wooden shacks built for the Student Army Training Corps during World War I."

The land owned by the university is among the resources of the Permanent University Fund, an endowment that provides financial support to the University of Texas and Texas A&M University systems through oil, gas, sulfur, and water royalties, rentals on mineral and grazing leases, and other investments. The endowment was initiated in 1838 when the Congress of the

Republic of Texas set aside more than 221,000 acres of land to fund higher education. When the University of Texas opened in 1883, another million acres was added, mostly in West Texas. The land was of little value until the Santa Rita oil strike of 1923, but by the beginning of World War II, wells in twenty-three oil fields were adding $1 million a year in oil royalties to the fund. Santa Rita No. 1 was considered so important to the university that in 1940 the Texas State Historical Association had the drilling rig moved from its original site to the campus in Austin. The well produced oil until it was finally plugged in 1990.[53]

The major principals involved in the development of Santa Rita No. 1 were Rupert P. Ricker, Frank Pickrell, and Haymon Krupp. Ricker, an attorney whose family owned an unsuccessful ranch in Reagan County, was convinced that oil underlay the land, and with the assistance of others obtained drilling permits on university land in Reagan and three other counties shortly after his return from World War I. But a big obstacle was the more than $43,000 in rental fees that had to be paid before the lease permits could be issued. Ricker was at first unsuccessful in convincing other, more experienced oilmen to provide him with the necessary funds. However, after a chance meeting in 1919 with Pickrell, who had served under Ricker in San Antonio during the war, and Krupp, a successful merchant from El Paso who was traveling with Pickrell, Ricker offered to sell them the leasing rights to the land for $50,000. Pickrell and Krupp counteroffered to buy the rights for $2,500 and Ricker reluctantly sold.

Pickrell and Krupp then proceeded to raise the money necessary to pay the rental fees, principally through a bank loan, and the permits were issued. With some friends from New York, they formed the Texon Oil and Land Company as a Delaware Corporation, with Krupp as president and Pickrell as executive vice-president. Later, the corporation would form a spin-off company, Group No. 1, which financed the development of Santa Rita No. 1 through the sale of certificates of interest in the Reagan County land. Group No. 1 would not incorporate until 1923, after Santa Rita No. 1 was brought in.[54]

Clayton joined Texon and Group No. 1 about a year after Santa Rita No. 1 hit oil. As noted above, his first contact with the company was with geologists who were working in the vicinity of Clayton's highway project in Sanderson. Kurt de Kusser, A. M. (Jack) Hagan, and I. W. Keyes showed Clayton how they identified promising surface formations that might indicate the presence of oil. As a result of his exchanges with the geologists, Clayton decided to seek employment with the oil company.

"Engineering jobs were playing out and [I] had to move all the time and I had been rather a restless mover from one job to another in engineering, especially highwaying ... The first chance I got I went to Big Lake and talked to Frank Pickrell ... he needed an engineer to make [drilling] locations and next places of drilling tests [and] he put me to work."[55]

When Clayton joined the company in November of 1924, Texon was emerging from a difficult period in which its financial resources had been severely strained while it tried to develop the Reagan County oilfield that had produced Santa Rita No. 1. By the summer of 1923, Texon had also begun drilling Santa Rita Nos. 2 and 3, further depleting the company's finances. Despite the promise offered by Santa Rita No. 1, Pickrell was unable to interest any of the major oil companies in joining Texon to develop the field, so he turned to Michael L. Benedum, a well-known wildcatter who had drilled successful wells in several states. After receiving positive reports from geologists on the potential of the Reagan field, Benedum entered into an agreement in which he received sixteen sections of land near Santa Rita No. 1 that would be developed by a new company, the Big Lake Oil Company. In exchange, Texon received cash and a guarantee that its expenses incurred in drilling wells Nos. 2 and 3 would be covered by Benedum. Texon also retained its interest in other sections of land around Santa Rita, which would continue to be developed by Texon's Group No. 1.[56] The foundation was now laid for Clayton's education as an oil field geologist.

CHAPTER 7

"Dangerous Business and Dangerous Surroundings"

Clayton's time in the oilfields from 1924 to 1928 mirrored the rough-and-tumble life of oilmen everywhere, and his stories illustrated the frontier lifestyle of the oil industry of that period, which was frequently marked by drunken fights that sometimes resulted in injury and death. "Dangerous business and dangerous surroundings—that was the code of the day," Clayton said years later. One story he told that highlighted the dangers associated with life in the oilfields began with an innocent dance at the Santa Rita Hotel in the town of Best.

A Group No. 1 mechanic named Blackie took umbrage when the wife of the town's deputy sheriff refused his request to dance. The deputy sheriff and C. A. Jones, a rig builder, threw Blackie out of the dance hall. Clayton said Blackie returned later with a revolver and shot the deputy sheriff in the ear, then turned the revolver on Jones and shot him in the arm, breaking one of his bones. "Jones took two long steps and was out the door, with Blackie after him. A girl who worked in the Best telephone office was standing just outside the door as Blackie came out. She screamed and Blackie turned and fired upon her. After she fell, he stood over her and emptied the gun."

The deputy and the young woman were killed, but Blackie lost Jones's trail. A day or two later Blackie was found in his own house, where he had presumably committed suicide. "Mr. Hayes [the deputy] was a fine man and officer, but there were times when law and order were not so represented and things got to such a state that the oil companies were forced to call in outside assistance for a cleanup," Clayton said. "The town that had the Best name was the worst one in Texas."[1]

In another version of this story, Clayton said that he and Carl Cromwell, Texon's chief driller and the man who brought in Santa Rita No. 1, were in San Angelo at the time of the shooting. Cromwell's foreman called and asked him to make arrangements for a doctor to work on Jones's arm.

Clayton and Cromwell made the 80-mile trip from San Angelo west to Best, located between Big Lake and Rankin. "We got a doctor, but all the doctors were out of town except a baby doctor," said Clayton. "While they were getting that arm bone straightened—trying to get the pieces of bone out of there—Jones would groan a little and Cromwell would say, 'Yeah, we got the right kind of doctor for you, you big baby.' That kind of cheered me up."[2]

Disputes in the oilfields were often settled with fists rather than words. Clayton was a party to such a dispute in 1926 involving a Texon production superintendent named Earl Willoughby, who apparently became jealous of Clayton's rapid rise in the company. "On one occasion he had talked pretty severely to me and in a threatening manner while carrying a hatchet in one hand," Clayton recalled. "At another time, he threatened me while he had a pinch bar [similar to a crowbar] in his hand." Fearing for his personal safety, Clayton bought a revolver and carried it on his person, "not with any intent of killing my opponent but to prevent him from hitting me with any weapon." According to O. W., who told the story to J. C., Willoughby was physically much bigger than Clayton.[3]

Matters came to a head when Clayton learned that Willoughby had "brow-beat" one of Clayton's employees. Clayton told Cromwell about the incident and said he was going to settle the matter with the foreman. Cromwell replied, "Good luck, go ahead." Clayton located Willoughby in the warehouse adjoining the Santa Rita railroad station and cursed him "in my most violent language." The foreman charged and Clayton punched him in the face, cutting the man's cheek. The two men wrestled for a while until Clayton succeeded in pinning Willoughby underneath him and twisted his arm behind his back. "I fully intended to break his arm, but the onlookers pulled me off from him. . . . I noticed then that he was bleeding from the lick I had given him."

Clayton invited the foreman to continue the fight outside but Willoughby refused, perhaps because he saw that Clayton was carrying the revolver under his belt. "I had completely forgot about that gun and it was quite likely that my opponent could easily have gotten that gun and killed me with it during the altercation. Under those circumstances, it had been foolish to have had that gun on." Grandson Clay Pollard remembers his grandfather's saying that it would have been better to have not been wearing a sidearm and to have taken the short end of a fight rather than the short end of a bullet.[4]

The fight finally put a stop to Willoughby's harassment of Clayton and his employees. And, to Clayton's amusement, the foreman, who had earlier bragged that he was "going to whip hell out of an engineer," told the doctor treating his wound in Best that a piece of machinery had hit him in the face.[5]

From an early age, Clayton was able to take on bigger and stronger opponents with his fists and win. Clay and Scott Pollard both recall that their grandfather was an excellent boxer and taught them how to fight with their hands. "Paw Paw said that if a man starts throwing quick punches at you, take a step back and let his momentum carry him into you, exposing his head to a punch." Scott remembers that his grandfather had shown him how to hold his fists so the knuckles were exposed, increasing the chances of landing a damaging blow. "He taught me a couple of tricks about boxing that I actually used when I ended up in an intramural boxing tournament at the University of Oklahoma. I ended up winning the tournament."[6]

Adam Pollard says that his grandfather taught him how to defend himself at age six. "With three older brothers already taught, I was the last Pollard boy to learn boxing from him. He showed me how to stand strong, control myself, and watch my opponent's reactions. The skills he taught me have served me well not only in the ring but in life. . . . He taught me not to be a bully but also never to be the one bullied. I fought in the marines using the same tactics he taught me as a child."[7]

Adam says that his boxing skills made an impression on his drill instructors. "I had been advised before going off to boot camp to stay unnoticed, just like a fly on a wall. My drill instructor didn't think I was tough enough to be a marine and put me in a boxing ring with a gangbanger from Los Angeles who was about my size. It was an act of disrespect to the gangbanger just to have to fight me, but I'm sure he thought it would give him some practice. I knocked him out with the first punch, so the drill instructor put in another guy and I broke his jaw. Soon all the other drill instructors gathered to watch. I fought six times that day against recruits from other companies in my battalion. I won every match and I was scared to death. Here I was a, small town Texas kid fighting gangbangers from LA and street kids from New York and Chicago. I can still remember how every fight went and I used the tricks that Paw Paw taught me. From that day on I was no longer a wallflower at boot camp!"[8]

While violence was a part of life in the oilfields, daily life was also highlighted by comical incidents. Clayton delighted in telling funny stories. One involved Joe Hoffer, purchasing agent for Group No. 1, who was fond of wine at a time when alcohol of any kind was forbidden in the camp. Hoffer had a recipe for fig wine and ordered the figs from California. He made the wine, but it was so strong that several men in the camp became drunk and incapacitated. The mash that was used to make the wine was thrown out to be picked up as garbage.

An old bull had the habit of making the rounds ahead of the garbage men. . . . The bull got a fair meal off of the mash. When Cooper [possibly S. O. Cooper, who represented a manufacturer of steel drilling derricks] came in he was chased by the bull.[9] Cooper dashed behind a chicken-net screen that had been placed for a back screen for a tennis court. The bull did not see the net and got tangled up in it, to the satisfaction of Cooper. It took considerable time to separate the bull from the net without anyone having personal contact with him.

In the morning the bull was still running loose, but he was yet under the influence of the mash. He was very dusty as he must have gotten down several times during the night. He appeared in front of Carl Cromwell's house where he was met with a broadside charge of rock salt. After the discharge of the gun, nothing could be seen in the direction of the bull except a great cloud of dust which arose out of his hide. When the dust cleared away, the bull could be seen about a quarter of a mile away, running for his life.[10]

Clayton also loved a story about a fight between Cromwell and Jones in an airplane being flown by the famous aviator Wiley Post. Post was a native Texan who had once worked in the oilfields and was the personal pilot for a wealthy Oklahoman. The men were returning from a big oilman's celebration in San Angelo where considerable alcohol was served.

"One was seated behind the other in the open plane," recalled Clayton. "Due to roughness and an over-indulgence in alcohol, the one in front was vomiting and thereby gravely offending the other. They were both immense men and when they began to tussle, the pilot nearly lost control of the ship. He brought them down as quickly as possible and informed them they would have to walk. This ended the struggle. Many times these two big men kept me from getting whipped just because of their size."[11] In another telling of this story, Clayton said that "Post landed the two men about five miles out on the road to Big Lake and told them, 'You bastards can walk now and tussle on the way.' I asked Cromwell about it and he said it wasn't such a long walk."[12]

Clayton once shared a hotel room in McCamey with the burly pair, who each weighed more than 200 pounds. McCamey was a boomtown established on the edge of the McCamey oil field in Rankin County. It was a rowdy Saturday night, and Clayton found Cromwell and Jones gambling in the Burleson Hotel. He had noticed their automobiles parked in front of the hotel while passing through McCamey with his brother Waldo. Waldo left Clayton at the hotel to spend the night with Cromwell and Jones. Since it was "a Saturday night and a time most men were heading to join their families, someone chided Cromwell about it was time for him to go back to his wife and family. He replied that some men were mice, but that he was the boss of the family. This matter he emphasized several times and nobody doubted the statement the first time he said it."

Clayton spent the night in Cromwell and Jones's room. In the morning, Jones took over the bathroom to shave and Cromwell left to find a barbershop where he could get a shave. As Clayton closed the hotel room door behind Cromwell, he heard footsteps in the hallway.

> Upon looking down the hall I saw a nice-looking young lady walking toward me, with her hand on her head as if she had a headache. When she said, "I'm sick and dizzy. Can I lie down a minute?" "Sure!" I said, as I directed her to one of our beds. She lay down with her clothes on and closed her eyes.
>
> I was standing in the doorway, wondering what to do for an unknown sick or dizzy young lady, when I heard another footstep in the hall. Upon turning around, I saw Mrs. Cromwell who appeared to be upset but not sick. She advanced upon and past me as she demanded to know where Carl was. I told her that he had gone down to get a shave. She then asked, while looking down at that girl on the bed, "Who is that?" I replied, "I don't know. I never saw that girl before in my life. She just now came here saying she was sick and dizzy and wanted to lie down a few minutes." By this time, I was myself getting a little dizzy as that young lady lay there on the bed with her eyes closed as if she were sound asleep with her clothes on. In the meantime, my good pal Jones had quietly closed the bathroom door and excluded himself from any involvement in the scene.
>
> Completely perplexed, I finally suggested to Mrs. Cromwell that she look for Carl at the town barbershop and I would look around the hotel. With that, she took off and down the stairs to enter her Lincoln car and follow my suggestion. Her car had no sooner roared off than Carl showed up, gathered up his gear, and we went down to check out of the hotel—as Jones reminded Carl that he surely was the Boss of his family or he would not be hurrying off to go home. As we got outside the hotel Mrs. Cromwell drove up, slammed on the brakes, and exclaimed "Carl." With that, his shoulders dropped as he said "Yes'm." I knew then, without any doubt, who the Boss was.[13]

In addition to being big men, Cromwell and Jones were toughened by the hard work in the fields. On one occasion, Cromwell's physicality kept Clayton out of deep trouble. "One of the large owners of stock in the Texon Oil & Land Company had considered and was contemplating a lease in eastern Pecos County. Pickrell sent me around with him to look at the region. The man who had the lease for sale did not have a bad reputation but was known as a gambler who had killed one man, and the people of Sheffield [Texas] had the saying that a funny thing happened there: 'the only witness to that killing, a Mexican woman, shot herself to death with a pistol.'" In another version of the story, Clayton said that "Sheffield was thereafter known as the only town in Texas where a Mexican woman had committed suicide twice through the heart with a sixshooter."[14]

Clayton said he advised the stockholder not to purchase the large lease. Several weeks later Clayton and Cromwell were attending a party in San Angelo when the man who was trying to sell the lease cornered Clayton and started threatening him. "My friend Carl Cromwell walked up. The lease broker walked away and didn't bother me again. I afterward learned or was told that when Cromwell worked on the drilling of the 'Trap Shooter Reilly' test, about a mile distant from Sheffield, he had a run-in over a poker game with this same lease broker. Cromwell knocked him down and then the broker said, 'I'll go get my gun' and Cromwell challenged him to do so. That ended the affair. After hearing that story, I kind of knew why Cromwell's arrival caused an easement of my situation."[15]

Clayton became good friends with Cromwell and Jones. In 1926, Jones allowed Clayton and Cromwell each to buy an interest in 160 acres of oil leases on university lands in the Church-Fields oil field in Crane County. Jones had acquired the leases in exchange for building a derrick at another location in the field. The deal ultimately brought Clayton $110,000 (equivalent to more than $1.3 million in 2009) when the group sold its interests to Frank Pickrell in 1927. Clayton received for his interest $10,000 in cash, $30,000 in Texon stock and $70,000 taken from Pickrell's interest in future oil production from the acreage. Clayton later said that he was soon paid out in full after Pickrell sold the lease and the stored oil production to the Warner Quinlan Oil Company.[16] The infusion of funds probably influenced Clayton's decision to leave Texon in 1928.

Geology in the Oil Fields

Clayton was hired by Pickrell to be the chief engineer for Texon and the Group No. 1 Oil Company, and he laid out the company town at Santa Rita and supervised the construction of the lighting plant, diesel engine power plant, and sewage system. The towns at Santa Rita and Texon "were built as model oil field camps to accommodate only their employees," he later wrote. The towns contained cottages for the married couples, bunk houses for the single men, restaurants, a main office, a small school house, loading racks next to the railroad, hospitals, independent electrical power plants, sewage-handling facilities, waterworks, telephone systems, garages, streets, and culverts. "It was my job to superintend all of this and its maintenance."[17]

Clayton believed that Pickrell hired him because Clayton had worked for the Chino Copper Company, which operated the Santa Rita copper mine in New Mexico. Pickrell, a gambler at heart, thought it was a good omen that Clayton had come from one Santa Rita to another.[18] Clayton's hard work was soon rewarded. "They gave me some stock in that company about two or three months after I had been working for them. Raised my salary. I [gave] them a hand there all the time—day or night—to do anything that needed doing. They furnished me a car, a Ford Coupe. A coupe didn't cost much in those days. I think $300 or $400 to buy. They also bought me a Chrysler just like they did Cromwell.[19]

Clayton moved up rapidly in the Texon and Group No. 1 ranks, and it wasn't long until he headed the geology and land departments for the companies, with responsibility for locating potential new drilling sites in Reagan County despite having had no professional training as a geologist. "I suggested they get a geologist," Clayton recalled later, "but they didn't want any part of a geologist. I suited them just right." Clayton's son Claytie says that his father had real talent as a petroleum engineer and became one of the first registered professional petroleum engineers in Texas. The legislation creating registration was passed in Texas in 1937, and Clayton is listed in the 1938–39 roster book of the Texas Board of Professional Engineers as No. 2651.[20]

Clayton began his education in oil field geology through his encounters with the Texon employees working in the area near Sanderson and expanded on that knowledge when he was befriended by Tom Prettyman, a geologist working for The Texas Company, later known as Texaco. Clayton also read reference books on geology. He learned to use an instrument called an alidade, which was mounted on a table with folding legs (a "plane table") and "shot" the elevations of the exposed features on the surface before transferring those data to contour maps.

"The plane table was used with the Alidade to determine distances, locations and elevations of key points in the strata along with the location of survey points on the ground," Clayton said. "By correlating the different key points and noticing the dip of the strata and bringing all of the values to one key point, the different points of identification could be mapped as a structural contour map."[21] The locations of existing water and oil wells were also transferred to the map, and the well cuttings were examined to determine subsurface formations at various depths. Clayton learned how to use acid on the cuttings to determine the depth and location of the Permian dolomite formations, what the drillers called the "Big Lime," in which oil production sometimes occurred. Dolomite, like limestone, is a carbonate, which is home to 40 percent of the world's petroleum. Clayton analyzed cuttings through a microscope to see if they were anhydrite or a form of gypsum and determined the top of salt formations by tasting the cuttings.[22]

Geologists at that time also used devices such as the gravimeter and the magnetometer to locate and map potential oil-bearing structures below the surface, although Clayton did not mention that he used either of these while working for Texon.[23] "I shall start our substructure maps tomorrow—1st on the pay-top—later on salts and red beds," Clayton wrote his father on February 24, 1925. The reference to salts and red beds was to various subsurface salt and sandstone formations that were passed through during the drilling. The pay-top or "pay" was the formation in which oil was encountered. "As soon as I get on to the oil end of it [I] expect to look into potash. I am taking samples every 10′ in salts."[24]

The presence of salt beds was extremely important to geologists as salt has been connected with petroleum since the Spindletop well was drilled through a salt-dome formation in 1901. While not all salt domes trap oil, there are many oil-producing formations along the Texas Gulf Coast and elsewhere throughout the world that are associated with salt domes. They are created when salt flows under pressure toward the surface in blobs or domes, breaking the overlaying strata, which forms a cap over the dome. As salt is impervious, it can trap any oil that the strata may hold.[25]

Learning petroleum geology "on the fly" as Clayton did was not unusual in the West Texas oilfields. As the number of wells increased, there were simply not enough professional geologists available to examine the well cuttings, and men like Clayton trained themselves to identify subsurface formations from the rock samples brought to the surface. Other self-trained geologists came from the ranks of the "oil scouts," men who worked for the oil companies and traveled the countryside in search of information about promising wells and drilling sites.

"Many scouts could tell the characteristics of drilling samples by looking at them," wrote oil scout Clarence Pope, "by watching and listening to the rhythm of the walking beam [a key component of a cable drilling rig], and then by suddenly hearing the sound change—interrupted and changed to jerks which shook the rig from top to bottom and signified that the tools were stuck in a crevice or cave-in." Cable-tool drilling dates back several thousand years. A cable drilling rig consists of a sharpened heavy metal bit that is attached by a wire line or cable to a beam above the rig floor that raises and lowers the bit in the drill hole. The bit falls to the bottom of the hole and hammers the rock layers.[26]

"Every oil scout has smelled hundreds of samples and tasted them—and what a thrill to hear them say, 'Sure smells like oil to me,' or 'No sulphur water here yet,' or 'The formation's sandy, though; I can tell by chewing a little of several of the samples,' or 'You can tell by the worn and out-of-gauge bit that we are drilling hard, sandy rock.'" Pope added that some scouts would carry their own acid with which to test samples to see if they were lime or gypsum.[27]

While both formally-trained and self-made geologists used proven instruments and their own training and experience to search for promising oil structures, countless others professed to have either a natural talent or a special device for finding oil, natural gas, minerals, or potential oil-producing structures. A popular name given to these people was "doodlebug," and O. W. encountered several during the hunt for oil in and around Fort Stockton.

"There are 2 'Doodle Bugs' now at work," he wrote J. C. early in 1926, "and they are continually setting up for some kind of measurement, first on one side and then on another side of the 7 Mile Hill, which leads us to suspect that they have detected some signs of a fault, or of a salt dome in that neighborhood, for it is the special merit claimed for the 'Torsion Balance' that it detects faults and salt fields."[28] The torsion balance was a legitimate instrument that measured fluctuations in the earth's gravity and was a predecessor of the gravimeter.[29]

According to Robert O. Anderson, author of *Fundamentals of Petroleum Geology*, some doodlebugs "claimed to find oil through the special properties of willow wands, electricity in their bodies, or ectoplasm. In the early days, when prospecting was mostly guesswork, anything that promised an advantage got a hearing, and doodlebugs thrived. Then, their record was probably as good as any other; today, it is poor."[30]

Clayton worked hard to learn the basics of oil field geology and still perform his engineering duties competently, and the heavy workload prevented him from visiting his parents in Fort Stockton, as O. W. observed in a letter to J. C. "Clayton is kept quite busy in his new occupation, so busy that he has not been home on a visit since he went to Big Lake. He has been lining up and constructing big storage tanks of 55,000 & 80,000 barrels, running [elevation] levels between the various wells, and trying to determine from these levels and the logs of the various wells how the underground strata 'dip' and 'strike.'" Well logs were a diary-like account of everything that happened or was encountered while drilling the well. "Dip" and "strike" are geologic terms that describe the orientation of tilted layers of rock beneath the surface, which was determined using a Brunton Compass. The compass could detect changes in the earth's magnetic field, helping geologists measure the dip of beds beneath the surface.[31]

Occasionally, the company used special drills to bring up solid cores of rock rather than chips, which allowed Clayton to see the thickness and relationship of the sub-surface strata. O. W. commented on this technique after observing his son give an orientation talk in 1926 about the subsurface formations at Big Lake. "The long strings of glass cylinders containing samples of the rock, taken regularly all the way down in the numerous wells of the Big Lake District, [were] hung from the high walls [of Clayton's office] side by side in such a way as to show all the undulations of the strata in the 2000 acres of drilled field [and] were appealed to for illustration."[32]

A Family Interest in Oil

Despite his customary caution, it is not surprising that O. W. shared his son's interest in the oil business. As a knowledgeable and respected lawyer, O. W. did the legal work on numerous oil leases for others, and he was at the center of the information pipeline about oil developments in Pecos and surrounding counties. His knowledge and experience led him to begin acquiring oil leases for himself and his family, and he followed developments in exploration near his land with keen interest. He also shared with Clayton a fascination with the new technology for finding and extracting oil, and he learned everything he could about it. Thus, it was with great expectation of future profits that O.W. wrote to J. C. in 1926 about the discovery of oil on the Ira Yates ranch located on the west bank of the Pecos River in southeast Pecos County. He related to J. C. the expressions of envy of colleagues in other counties: "It has been said that there was neither oil nor law west of the Pecos; now you have both." The Yates field became one of the most prolific oil fields in the United States, producing 1.4 billion barrels of oil through 2006.[33]

Both Clayton and O. W. had connections to the vast Yates field, although neither benefited directly from the discovery of oil there. Shortly after the Transcontinental Oil Company drilled the discovery well on the Yates ranch in 1926, Texon directed Clayton to map the Yates field. He dispatched Waldo and a crew to the area, and the map, which proved remarkably accurate, included an extension of the field to the west that had been missed by other surface maps.[34] Many years later, O. W. revealed to J. C. that in 1893 he had surveyed the land that included the Yates field and had acquired two tracts that he later sold for twenty-five cents an acre. "I am told that on one of these tracts a 200,000 barrel well was brought in, and that the 2 tracts are valued in the millions. I can not say however that I grieve any over it. It may have happened to others, and 40 years is a long time to wait even for a fortune."[35]

The Yates oil boom improved O. W.'s law practice and leasing business through both the sale of leases and the resulting increase in legal work. "The oil boom [reveals] weak spots in the titles to many lands, when placed under the fire of some Oil Company lawyer, and then we get considerable income from buying and selling oil leases," O. W. wrote J. C. in the spring of 1927. "My income this last year was nearly $8,000 [equivalent to about $97,000 in 2009] and promises to be greater this year. . . . It was a brighter year for the family than for years back, and even now the only dark spot on the family horizon is your situation—tied up in China."[36]

While O. W. was thankful for the increase in income, he despaired at how the influx of oilmen and oil scouts, speculators, and "curious travelers" was changing the complexion of his small West Texas town. "When I walk the streets at least half of those whom I meet are unknown to me, and I can not call at once the names of half of those to whom I speak," he wrote J. C. in 1927. "The public schools will be crowded to the limit, and the Courts docket is congested with suits, most of them concerned directly or indirectly with oil leases. When you get here you will come to a town that is as new to you as Shanghai or Hankow."[37]

When his work permitted, O. W. would accompany Clayton on visits to drilling sites near Fort Stockton. On one such visit in 1926, O. W., his wife, and a family friend rode to the site in Pickrell's Chrysler with Clayton at the wheel: "The car," O. W. wrote,

> was capable of running 60 to 70 miles an hour. We were leaving late, about 5 P.M. The road was good (Old Spanish Trail), and over much of it we made 50 & 55 miles per hour. We turned off the Highway a mile before getting to the well—a rough mile—yet our time to the well—33 miles—was just 43 minutes. During that time your mother & Mrs. Cato must have seen very little of the country. Every time I looked into the back seat, I saw them stretched out, almost flat on their backs on the seat, which they said was the only position they could take to avoid a continual butting of the head against the top of the car. But they were game and when Clayton offered to slow up they told him to keep up his gait so they could boast afterwards of having traveled at the rate of a mile a minute. It was a John Gilpin ride, but we lost nothing but speech.[38]

In early February 1925, a sizeable natural gas well came in near the loading dock at the Orient railroad tracks in the Big Lake field. According to historian Myres, the well was uncontrolled and produced 60 million cubic feet of gas daily, and there was great concern that trains passing through the area could cause an explosion. Like so many other developments in the West Texas oil business, O. W. chronicled this event for J. C.

> The gas was totally unexpected & it is a mystery yet how it happened that the well did not take fire, so that some people have suspected the presence of helium. But great care is being taken while the effort is on to cap the well. It is close to the R.R. track and the trains are stopped & the order given "Lights Out" while the cars are carried through the danger zone by the initial drive from the engine to be carried on by another engine after passing the danger. Cigarettes are "tabu"—not allowed. We have heard nothing from Clayton yet about these matters although he must be in the danger bounds—because it is a well in which Pickrell is interested—and I think he does not want to make your mother uneasy.

The well was not capped for five days until special equipment and an expert arrived from Tulsa.[39]

Gas was also produced from oil wells, distinguishing it from gas produced from natural gas wells. This gas, sometimes called "casinghead gas," separated from the oil as the pressure was reduced at the wellhead, and it contained a variety of liquid hydrocarbons, including natural gasoline, a highly volatile substance. Casinghead gas was sometimes used as a fuel in the field but was normally flared. Clayton soon learned to use crude field tests to determine the presence of the gasoline and demonstrated one such test to his father at a wellhead.

"He said this gas was petroleum gas of the 'wet' class; that is, it carried gasoline while 'dry' gas carries none," wrote O. W. "His test of this is called the handkerchief test, because a handkerchief thrust into the escaping gas [carries] back with it the odor of gasoline.... I am told that a blotting pad used in same manner as a handkerchief gives an even better test, and in this way crude field tests are made to determine the character of gas when found in drilling."[40]

Frequently, when drilling failed to produce oil or natural gas in sufficient quantities, a technique was used to detonate nitroglycerine in the well hole to create fractures in the subsurface strata that would release what was below. O. W. was present when a well was "shot" and described the technique:

We waited at the rig quite a time before the gun-cotton was placed in the hole, and during that period autos continuously arrived until I counted 58 in all when the explosion came. [With care] the 'shooter' and his assistant brought the 'stuff' [that] was marked 'Nitro-Glycerin' and carried 10 cylindrical hollow tubes about 5 feet long, beveled to a point at one end and hollowed at the other so that the beveled end would fit nicely into another similar tube of like shape. These were called 'Cans' & seemed to be made of a light tin. When ready for action these cans were filled with water, then emptied, and smaller iron tubes containing the 'stuff' were somehow fitted into the 'Cans' and 5 of these 'Cans' containing the iron tubes—each loaded with 30 quarts of dynamite—were lowered to the bottom of the 2232′ hole.

. . . When ready the 'shooter' stepped out of my sight for a moment and a minute later I heard a sound like a fire-cracker going off at the mouth of the hole, but so light and weak that I did not realize that the shot had been fired until some 10 seconds later when—with quite a roar—a column of black oil rose 100 feet in the air, leaned with the wind, and came down all the way to the rail-road track 100 yards away, with a splash and a thump now and then as a small rock would strike the ground. After about 25 seconds or more this ceased, quiet came, and all that was in evidence was a derrick dripping with oil and a black band 30′ wide from the rig to the railroad bed. It was not a gusher but a 'pumper,' estimated at 50 to 100 barrels. . . . Whether this would endure remains to be proven.[41]

During his years with Texon, Clayton did the surface geology necessary to define the hypothetical lines of the Reagan County field and to determine promising locations for Group No. 1 to drill. By the spring of 1928, Group No. 1 had 78 wells that were producing a total daily average of 5,642 barrels of oil from the Big Lake field.[42] "I never drilled a dry hole," Clayton said proudly.

Chapter 8

Drilling the University No. 1-B Well

About two years after joining Texon, Clayton began to suspect that oil deposits might be found at depths far greater than those penetrated by the successful wells previously drilled by Texon and Group No. 1 in the Big Lake field. The first clues came when he analyzed drill cuttings taken from the so-called "Mystery" well, which, at 6,004 feet, was the deepest well sunk at the time (Santa Rita No. 1 had struck oil at a depth of 3,050 feet).[1] The well was drilled by Group No. 2 Oil Corporation, another spin-off of Texon that was incorporated in 1921 principally to drill for potash deposits. Potash is a key ingredient of many fertilizers, and in the 1920s, as today, most potash used in the United States was imported. Finding a commercial quantity of potash would be almost as valuable as a significant deposit of oil.

Unfortunately, the Mystery well found neither potash nor oil.[2] Clayton sent the drill cuttings to the University of Texas for analysis and learned that there were ammonia fumes in a zone of bituminous shale located beneath the "Big Lime" horizon. "I had been reading recently on the oil shale industry in the British Isles and had noted that ammonia was frequently found with these oil shales," Clayton wrote later. "Whether there [was] any relation of significance [was] questionable, but it was a germ of suspicion."[3]

The Mystery well had been drilled on the northeast edge of the Big Lake field, about one mile from Group No. 1's productive wells. Clayton analyzed the surface structures in the area and compared them with a survey of some exposed geologic formations in the Glass Mountains of Pecos County about a hundred miles southwest of the Big Lake field. The survey was provided by a geologist with Humphrey's Oil Corporation. "After some talk and comparison, without anything but a hazardous guess, we made an estimate that the Pennsylvanian might be reached at around 6,200 feet," Clayton later wrote. The Pennsylvanian is a period of the Paleozoic era when great coal deposits were formed, approximately 318 to 299 million years ago.[4]

However, Clayton needed more evidence that there was oil at the deeper levels, for Group No. 2 had expended considerable resources on three wells that had come up dry on the eastern edge of the field, and management did not want another dry hole. By analyzing the subsurface contours of Group No. 1 wells that had struck oil at the Big Lime horizon of roughly 3,000 feet, Clayton hypothesized that fault lines might be allowing the upward migration of oil and gas from deeper formations to the shallow formations. He had crews drill into the field east of the Group No. 2 wells to examine the sand section of the Big Lime horizon there. "We had no equipment available [on site] to examine these samples, but it appeared to me that the sample contained small particles of hydrocarbons. This, if true, might be an indication that oil had traveled through this formation, leaving these particles caught in the sand, while the volatile matter continues to move upward due to water pressure. This pretty well covers the question of why I thought we might have deeper pay." He also noticed that as the location of the producing wells moved east, the depth at which production occurred increased from 2,500 to 3,000 feet, and that the 2,500 foot wells had heavier oil and more natural gas, lending some credence to his hypothesis about the upward migration of oil in this area. For these reasons, Clayton decided to locate the University 1-B well about two miles east of Santa Rita No. 1, just south of the Kansas City, Mexico and Orient Railroad tracks in an area surrounded by producing wells.[5]

But before he could begin drilling on the site, Clayton had to convince Pickrell and Cromwell that there might be oil at the greater depth and that it would be worth the expense to reach it. Texon would use the cable-tool drilling method to reach the oil. Due to the depth of the oil as hypothesized by Clayton, the project would be long and expensive. Cable drills were effective in smashing through rock formations to reach oil and gas at relatively shallow levels, but to reach oil at this depth, the drilling could take several years. Rotary drills during this period, though faster, were used to drill through the softer formations on the Gulf Coast. Eventually—by World War II—rotary drilling supplanted cable rigs due to technological innovations and the need to drill deeper holes, but that option was not available to Group No. 1 in 1926.[6]

Clayton's argument that there could be oil at deeper levels was convincing, and Pickrell enlisted the financial support of Levi Smith, president of Big Lake Oil Company, to pay 50 percent of the costs of the deep test well. Cromwell began drilling on February 8, 1926, but after the drill had reached 1,000 feet, Smith pulled out of the project and refused to pay any of the costs, stating that his geologists in Pittsburgh disagreed with the location. "Evidently there was considerable doubt on the part of the Pittsburgh office as to the advisability of the location," Clayton wrote later. "I was never advised as to what it was, but I am certain it was pretty fair reasoning on their part."[7] Because Group No. 1 was bearing all the costs of the test, it decided to recover any oil that was found as the drill went down. By the middle of August 1926, more than 32,000 barrels of oil and 400,000 cubic feet of natural gas were recovered, revenue that was sorely needed, as the cost of the well ultimately totaled more than $135,000 (equivalent to more than $1.6 million in 2009).[8]

The story of the difficulties encountered in drilling the University 1-B well has been well documented by Myres and others. Here are the high points as provided by Myres and Clayton:

- At 5,315 feet the rig caught fire and everything was destroyed except the steel derrick. Cromwell had the rig rebuilt and resumed drilling.
- At 6,985 feet the steam engine lifting the cable and tools gave out. It was replaced with electric motors and drilling resumed.
- At 7,535 feet, after two years of drilling, the steel cable was replaced. At 7,955 feet, the cable broke and the tools were lost at the bottom of the hole. Efforts to recover the tools failed and a second set of tools was lost. It took eight days to locate and remove the two sets of tools and resume drilling. Drilling tools frequently became detached and lost in well holes and had to be recovered or "fished" out using special tools designed for that purpose.[9]
- At 8,000 feet the electric motor burned out. It was replaced and drilling resumed.
- On the morning of May 1, 1928, fire broke out on the drill rig again. "I followed our fire-fighting brigade to the deep test to find that I was in charge of putting out an oil field fire, with no previous experience and nothing but some small foamite cans to work with," Clayton wrote. With help from other men, Clayton was able to save the derrick, but the rig and cable were destroyed, necessitating delays before drilling could restart in August.[10]

When the University 1-B well finally struck oil at 8,525 feet on December 1, 1928, Clayton was not there to see proven his hypothesis that commercial quantities of oil did exist at deeper levels in the Permian Basin. He had resigned from Texon on August 1, having become engaged to be married. He would later say that he resigned because he did not want to raise a family in the rough world of the oilfields. Clayton had hired his brother Waldo as an assistant engineer in 1926, and Waldo was on the site during the last few months of drilling.[11]

During this period, Waldo observed two instances in which he suspected that someone was trying to sabotage the project. One incident involved a field superintendent whom Waldo suspected of trying to "queer" the deep well because of a dislike for Clayton, using a technique to fish a cable off the bottom of the hole. This might have caused damage had not Cromwell stepped in. Another incident involved finding parts of a combination wrench that Waldo suspected had been put in the drill hole at about 6,700 feet for spite. "At any rate," wrote Waldo later, "whether accidental or purposeful, it had caused us a big expense and trouble."[12]

The story of the completion of the University 1-B well is a part of Texas oil history lore. According to both Myres and Waldo, Pickrell decided that enough money had been spent on the project and ordered Cromwell to stop drilling at 8,500 feet. "Carl accepted the order," Waldo wrote later, "but later when talking over the prospects of a producer with me, he suggested that he disappear without giving me the order to shut down and that I carry on down until we had a producer or a dry hole. On December 1st, the bailer [a bucket-like piece of equipment used to remove mud and rock cuttings from the bore hole] would not go down against the gas pressure and on the 2nd of December the well flowed forty barrels of high gravity oil."[13]

It took Waldo and the crew two days to locate Cromwell, who was drinking heavily in Sweetwater, Texas. "However, he was sober the minute he absorbed our message and within five hours was on the ground in efficient charge of the situation."[14] By January 1, 1929, University 1-B was producing more than 1,600 barrels a day of high quality oil and, according to Myres, "proved the practicability of deep wells both in the Big Lake field and elsewhere in West Texas."[15] The University 1-B was the deepest producing well in the world at that time and was the first production from Ellenburger (Ordovician) limestone in the United States. "The fact that the Texon group reached this depth, using outmoded cable tools, with a very favorable outcome, remains one of the outstanding accomplishments in the entire history of the petroleum industry," wrote Myres. Additional drilling produced 32 million barrels from the Permian Basin Ellenburger pool over the next twenty-five years.[16]

Locating the University 1-B was the most notable achievement in Clayton's oil field career and would provide the foundation for his reputation as one of the Permian Basin's significant petroleum pioneers. He did not boast about the discovery in his personal papers and letters, writing only about specific details associated with the location and drilling of the well. But he was fiercely proud of the discovery, Janet Pollard says. "I remember one Christmas, after I was married, we were in Fort Stockton at Daddy's house and a man dropped by. This man had recently had some success in the oilfields, and he was talking about various wells, including the University 1-B. 'I located that well,' I remember my dad said, and the man replied, 'The hell you did.' My dad got very angry, stood up, and said, 'The hell I didn't.' I thought he was going to get into a fight, but the man backed down and apologized."[17]

The Road to Marriage

Clayton worked hard at Sanderson and at Texon, but he still found time to have a social life. He loved to sing, played the piano, was a keen dancer, and drove many miles on barely passable roads to attend dances with young women from the area. Throughout 1923 and 1924 he carried on a romance with a "Miss Hanson," who was from Fort Stockton and was a student at the University of Southern California. He once made the long drive from Fort Stockton to Los Angeles to visit her at Christmas. However, by the end of the next year the romance was over, and Clayton told his father that Miss Hanson was engaged to be married.

Recovering quickly, Clayton was soon back on the dance floor with a young school teacher from Sanderson.[18] He continued to enjoy social engagements after moving to Texon, sometimes to the point of exhaustion. "There was a dance [in Fort Stockton] Friday night, and Clayton came all the way from the oil field to dance 3 hours, and then at 2 A.M. to pull back in his car some 75 miles to his headquarters," wrote O. W. in 1926. And in another letter describing late-night dancing and driving, O. W. observed that "young folks have forever ahead of them—or think they have—so waste of a little sleep is of no consequence—where a dance is in view."[19]

Clayton was now in his early thirties, and O. W. worried that his son was spending too much time socializing and needed to marry and settle down. "Clayton keeps quite busy but shows no symptoms of marrying," O. W. wrote J. C. "He will wait too long I fear." He cautioned the also unmarried J. C. that while he was working in China many of his "old flames" back home were marrying and that J. C. would have to find his future wife from a younger generation of women. "Clayton is even further behind than you are, and soon he must look for his future among the widows and old maids. Whatever his consolation may be, he does not seem to worry over the situation. Other matters give him more concern."[20] O. W. also discussed his concerns with Clayton, who told his father that he would not consider marrying until he was out of debt and had something with which to support a wife. "If he follows this idea out, he may not marry at all, or marry late, when he can hardly have as good a bunch to pick from as a younger man ordinarily would have," noted O. W.[21]

The young Chicora (Chic) Graham, whom Clayton met in the summer of 1927 and courted for a year before the couple married in San Angelo, beginning a partnership of nearly fifty-five years.

All of this changed in the summer of 1927 when Clayton met Chicora (Chic) Graham, the twenty-one-year-old sister of Oscar Graham Jr., who worked for Clayton at the Texon light plant in Santa Rita. Chic had finished her sophomore year at Trinity University, which was then located in Waxahachie, Texas, and was living with her family in Sterling City for the summer. With her mother Evie Lee (Mernie) Graham and friend Mike Moore, Chic embarked on a trip to the Big Bend country for a visit with her Uncle Joe Graham and Aunt Molly at their ranch on the Mexican side of the Rio Grande near Boquillas. On the way, the three stopped in Santa Rita to visit Oscar and deliver a chocolate cake, and Oscar introduced his boss to his mother and sister.[22]

"I thought she was the cutest little thing I ever saw," recalled Clayton. Chic thought Oscar's boss was "handsome." When Chic, Mernie, and Mike returned to Sterling City, Chic learned that Clayton had called. "When I got to talk [with him] he invited me to a dance at Santa Rita where I would stay in his brother Waldo and Olive's home. Mernie gave me permission. . . . Clayt came and took me to the dance and we had a great time. He was a wonderful dancer and even more handsome than I had thought. He had a little much to drink and executed the 'Buck and Wing Dance,' whatever that was! But we just had such a good time—drove over to Rankin for lunch at Big Lake and then he took me back to San Angelo [where Chic was visiting friends]. . . . then on the third date—he said he wanted me to marry him and we would go to Paris, France, for our honeymoon."[23]

Clayton chose Paris as a honeymoon location because he was planning to return there in September to attend the American Legion national convention and the celebration of the tenth anniversary of the signing of the Armistice. But Chic wanted to finish her college education before marrying and the engagement was put off. Clayton took a leave of absence from Texon and left for Paris by steamship on August 9, 1927. Chic's family moved to Alpine, and she enrolled that fall as a junior at Sul Ross State Normal College (now Sul Ross State University), majoring in English and minoring in history, with plans to finish her degree by the end of the summer term of 1928.[24]

Upon arriving in Paris, Clayton learned via telegraph that J. C. would arrive by train in Paris on September 1 from Harbin, China. Before meeting J. C., Clayton was able to squeeze in a few days' visit with the Gaubert family at their summer home in Saint Jean de Monts on the Atlantic coast southwest of Nantes. When he returned to Paris, the brothers toured the city for a few days before heading off on a trip by airplane to London, Amsterdam, and Brussels. They returned to Paris in time for the start of the convention on September 19.

While in Amsterdam, Clayton purchased a diamond "in contemplation of influencing a very pretty girl to become my wife." Clayton now knew that Chic was seeing someone else, but "thought that with some real continued effort on my part and with the aid of a fine blue diamond for an engagement ring, I might prevail upon her to change her affection to me and become my wife."[25] Mike Moore was the someone whom Chic was seeing. "You have heard, 'Absence makes the heart grow fonder'—of someone else," Chic told her children years later. "This was true for a kind of flighty girl." That fall Moore presented Chic with an engagement ring. "I accepted it and thought he was so nice but not as nice nor as handsome as Clayt," Chic wrote. "So I was engaged. But these beautiful letters began to come from Clayt (I have all of his letters to me) [and] a cablegram and several telegrams through September and October, and so in November that integrity that I believed I had won out; I gave Mike back the pretty ring."[26]

Clayton, 1927, in Paris, where he attended a convention of the American Legion. He had hoped to turn the trip into a honeymoon with Chic, but she put off the engagement to finish school. Clayton's brother J. C. came to Paris from China to visit, and the pair made a quick airplane trip to London, Amsterdam, and Brussels. In Amsterdam, Clayton bought a diamond for Chic.

Letters to Chic

Clayton's letters to Chic contained his usual detailed descriptions of people, places, and events he witnessed, combined with the longings of a man in love with a woman who was not quite sure who she wanted to settle down with. "You can tell I am constantly thinking of you," he wrote at the end of a long, detailed letter about his first day in Paris. "You have my curiosity aroused as to your loves—and I look forward with pleasure to your next letter."[27]

Peevishly (and sounding like his father), Clayton wrote at the end of his trip that he had received only two letters from Chic during his two months in Europe. "From August 28th on I never received a line from you—which covers a period of nearly 2 months. Now do you think that nice? I was depending on your letters as an indication of friendship or otherwise and from results I should judge that I lost out rather suddenly."[28] After receiving a letter from Chic upon his return to the United States, Clayton wrote back, "If you knew how happy it made me to receive letters from you I am sure you would either quit altogether or write more frequently."

Perhaps with a motive to make Chic jealous, Clayton described taking a date to the picture show in Pecos. "She was a very pretty girl—uses too much paint and powder—her mother owns the Pyote Hotel. We had a nice time." Apparently asked by Chic what occurred at the end of the date, Clayton confessed that he fell asleep in his car and did not wake up until 6 A.M. Aware of Chic's engagement to Moore, Clayton closed a letter with a plea. "I think you should change the last ring you have on your finger for the square shaped stone I brought from Holland. The square is the latest out and quite the most popular at the very present. I am afraid I have gone too far to love you a wee bit."[29]

By the end of November 1927, Chic had returned Moore's ring and spent Thanksgiving Day with the Williams family in Fort Stockton. Clayton, though overjoyed at the turn of events, was also philosophical about Chic's rejection of Moore. "My one big wish is: if I am ever in the same circumstances you will tell me—won't you, sweetheart? The unkindest cut one could give me would be the pretense of a continued love which did not exist. . . . I would dread such a situation and would miss the consolation and thoughts of my little dream cherry plum but better that than the bitter disappointment of an uncherished love. So my dear little sweetheart, you will always be as good to me as you were to him. Maybe perhaps I shall ask you to be better to me by telling me sooner in developments than you did him."[30]

But even after Chic's rejection of Moore, Clayton remained insecure about her feelings for him. After receiving a letter from Chic in December telling him that she would not have much time to spend with him during the Christmas season, Clayton responded with what he called a "bawling out." "I shall keep in *mind your letter* as I am sure you have not time; but you do not care if I drop by to see someone else in Alpine who possibly has just a little more time to spare—do you?? I know you are broadminded just as you would want me to be and *'you don't any more mind'* [Williams's emphasis]. . . . Under the circumstances it would appear that I had better make my own plans [for Christmas]." He closed the letter sadly. "You do not love me as I do [you]."[31]

The breach quickly healed, and sometime between Thanksgiving 1927 and January 1928 Chic accepted Clayton's offer of marriage and they began making plans.[32] However, the relationship continued to have its ups and downs. Learning in January that Chic had invited a male friend to visit her in Alpine, Clayton exclaimed, "I am trying to break you of the habit of dates except with me so when we are married it will not be so sudden a change—but at the same time I do not wish to cause you the loss of any great pleasure." The invitation had left "a bad taste," Clayton wrote.[33] However, upon hearing Chic's explanation that she did not want to hurt her friend's feelings, and after meeting and sizing up the young man as less than threatening, Clayton confessed the source of his jealousy. "Fortune told you nothing you did not know about your husband's jealousy. . . . I am afraid he will always be jealous—because he loves you so. He considers you a great prize whom others covet and [whom] he must guard—fuss—quarrel [over] to the utmost—that he alone may be sure of your love."[34]

No doubt contributing to Clayton's anxiety was his busy work schedule that limited his time with Chic. In addition to performing his duties as engineer and geologist at Texon, Clayton had convinced Pickrell and the Texon management to pump their oil wells with electric motors rather than motors powered by natural gas, and by early 1928 he was supervising the conversion. According to Clayton, the gas-powered pumps were frequently down due to hydrogen sulfide in the gas. The changeover was so successful that many other companies followed Texon's lead.[35] Clayton's business interests had also expanded beyond Texon. In anticipation of his future marriage and separation from the company, Clayton had invested in a water and ice plant in Crane City and in the Pecos Valley Oil Refinery at Pyote.[36] These investments, coupled with his grueling work schedule no doubt intensified his anxiety over the relationship with Chic.

"Sweetheart, I am tired to death," Clayton wrote in March 1928. "[I] would just like to sit back and listen to you. I hope it will not be long before I have that privilege."[37] As winter shifted into spring, his feelings of anxiety intensified. "Sweetheart, do you love me or do you not? Sometimes I think

JULY 2 1928 NO. 1 2

Clayton and Chic riding mules (last two in the group) at the Grand Canyon. It was in the summer of 1928, and they took the trip chaperoned by Chic's mother Mernie.

not. Be fair with yourself and me—do you sometimes and don't you other times—or are you sure all of the time?"[38] But after a reassuring response from Chic, Clayton's faith in her love was restored. "Now I felt like, at times, you did not love me on account of some little things you do or don't do & etc. But if I ever had any doubt your last letter must have squelched it."[39]

On September 10, 1928, two days after Chic graduated from Sul Ross, the couple was married in the Methodist parsonage in San Angelo. It was the beginning of a fifty-five-year partnership.

CHAPTER 9

Risk, Failure, and Reward

Having resigned from Texon and married at age thirty-three, Clayton now set about building a life that would revolve around the boom-and-bust cycles of the oil business without, as he would later put it, having to live in the sulfur stink of the oil fields. He wanted to be an independent oil operator with his home and headquarters in Fort Stockton. As noted earlier, Clayton took some of his profits from the Church-Fields oil deal with Cromwell and Jones and invested $15,000 in the Pecos Valley Oil Refinery in Pyote, about forty-five miles northwest of Fort Stockton. His other major investment was $55,000 in an ice plant and waterworks for the new town of Crane, located about midway between Odessa and Fort Stockton.[1] Unfortunately, neither of these investments would produce the financial results that Clayton wanted.

The Pecos Valley Oil Refinery was one of several small local refineries built to process some of the production from the Hendrick field, about fifteen miles north of Pyote, where oil had been discovered in 1926. The refinery could process 5,000 to 8,000 barrels of crude daily, producing gasoline, kerosene, lube, and fuel oil.[2] Clayton acquired a one-quarter interest in the refinery, and while initial dividends from the investment were good, by 1929 Clayton and his father had begun to suspect that something was amiss. The refinery incurred considerable debt while expanding to meet expected increased production from the Hendrick field, but the additional crude did not materialize, as oil producers cooperated to limit production to keep prices high and reduce water incursion in the wells. The curtailment, which was administered by the Texas Railroad Commission, was called "prorating" and was applied to production from both the Yates and Hendrick fields. Proration became a statewide issue in the early 1930s, when vast quantities of oil flooded the market from discoveries in the East Texas oilfields. By March of 1929 the Hendrick field was producing only six percent of its estimated potential, and producers were unable to live up to their production contracts with Clayton's refinery.[3]

On July 13, Clayton and O. W. were in Fort Worth to attend an emergency meeting of the refinery's board of directors. There they learned that the refinery's superintendent had borrowed $20,000 from one of the company's other stockholders to meet financial obligations and invested $150,000 in new equipment that was not being used because oil producers were not meeting their contractual obligations. The $20,000 note was now due and the company could not pay; furthermore, the refinery needed an additional $15,000 to meet other financial obligations. The stockholder holding the $20,000 note demanded to be put in charge of the company's management until the next spring to secure his note. If the board agreed to the request, the stockholder would put up the additional $15,000. The board acceded to the stockholder's request but was unable to save the company, and it went into receivership during the summer. Neither Clayton nor any of the other directors received a penny from the dissolution of the company, although one of the directors used the refinery as collateral to borrow several thousand dollars and then disappeared. Clayton said that Frank Pickrell had urged him not to invest in the refinery.[4]

Building a Home and a Business

The ice plant and waterworks that Clayton built in Crane provided him with much-needed income during the lean years of the late 1920s and early 1930s, while he was struggling to establish himself as an independent oil operator in the Great Depression. Built in 1927, the ten-ton plant was fed by a 500-foot water well, and the water was distilled by two boilers and a cooling tower. In addition to selling ice, the plant sold the water for a dollar a barrel.

A year later, Clayton noticed that the Phillips Petroleum Company had arranged to pipe water into its gasoline plant north of Crane. Obtaining a franchise and an easement from the Crane County Commissioners Court, Clayton built a pipeline from the Phillips plant to his ice plant, lowering his costs by substituting the soft water coming through the pipeline for the hard water from the well. He laid a few pipelines from his plant through the town. At its best, however, the ice plant sold only three tons of ice a day in the summer months, and during the winters it lost money. Household demand for ice was declining because improved design of home iceboxes meant they needed to be filled less frequently. More important, several years earlier Kelvinator had introduced the household mechanical refrigerator, and by 1935 there were 1.7 million home refrigerators in use in the United States. Only the sale of water kept the Crane plant open.[5]

A few months before his marriage, Clayton had begun construction of the fine white limestone house that he and Chic occupied in Fort Stockton for most of their married life. Clayton did much of the work himself, and the house, which still stands today, was built for $25,000, just down the street from O. W.'s home and office on the grounds of the historic fort. The construction process was not without excitement, Clayton recalled. "When I was building our house there was a fellow I kicked out because he was charging us to paint my father's office and then was using our paint on other jobs. I called him to come up to my house to chaw on it." Clayton found the man inside the house, walking with hobnail boots on a wood floor that had just been varnished. "He was just stomping around on it. I struck him in the hall and hit him not a very hard blow because if I had missed him my fist would have struck the wall. We decided to go outside and with the sun at my back I slapped him across the face and he ran at me. I stuck my fist right in his face and knocked his jaw out of place. I told him I could give him some more if he wanted it. He decided that was enough and went down and had the chiropractor put his jaw back in place."[6]

Janet Pollard said the painter had at one time been a good semipro boxer. "My father was pleased that he was able to knock the man out quickly with the first blow because he knew that the man would beat him in a longer fight. I remember my dad telling my brother that if you are going to be in a fight, hit first and hit hard and maybe the fight would go no further."[7]

Clayton and Chic moved into the house in February of 1929 while finishing touches were being made on the bathroom tile and fireplace. The design of the house reflected Williams's interest in French design and architecture acquired during his World War I service in France and his trips around Paris and the French countryside. Typically, O. W. had an opinion about the new house, terming it "so large . . . that I think at times they get lonesome. Chick [*sic*] is educating herself in cooking and house-keeping & Clayton is setting out trees, and sowing blue grass and flowers, and so they keep occupied during the day time—but the evenings must seem lengthy as neither one seems to have much taste for reading or studying up on anything but the current topics of the daily newspaper, and thus the books in their young library get dusty."[8] In his spare moments, O. W. would walk over to his son's house to water the trees, plants, and lawn. The judge's attention to taking care of the landscaping at Clayton's home would result in a tragic accident many years later that would mark the beginning of O. W.'s final decline.

Over the years, a long north-south hallway that ran through the house became the natural place to hang framed photos of friends and family. Clay Pollard remembered that anytime visitors came to the house, they would get a tour of the family photo gallery. "It was almost like being in a museum," he recalled. "I'm sure if you asked anyone who visited the house when Paw Paw and Grandma Chic lived in it, they would remember the hall of photos."[9]

Clayton worked hard to save the refinery and improve operations at the ice plant and water works, spending many hours traveling between Fort Stockton, Pyote, and Crane. He also pursued his goal of becoming an independent oil producer, and by the end of the decade Clayton and O. W., now even more enthused about the prospects for "local" oil, had invested in several thousand acres of oil leases in Pecos County. In 1929 Clayton began drilling the first of several exploratory wells, and the workload became so heavy that he and O. W. employed a clerk. Waldo joined the family business in September as a geologist and oil scout, having lost his job with Texon after the company was acquired by the Marland Oil Corporation. All of this activity required capital, and Clayton sold his Texon stock for $48,000 (more than $600,000 in 2009) to help finance the various ventures.[10]

Initially, Clayton focused his wildcat efforts in an area northeast of Fort Stockton then known due to its structure as the Fort Stockton "High." A "high" in geological terminology is the uppermost part of an inclined underground structure. The northern part of Pecos County contains the southern part of the Central Basin Platform, a major subsurface geologic structure running southeast from the southern portion of Lea County to northern Pecos and Crockett Counties. The southern edge of the platform is located north of Fort Stockton and contains a structure known as the Fort Stockton Anticline, which holds numerous oil and gas reservoirs. Williams had high hopes for his first wildcat well as it was located south of a producing well drilled by the Pecos Valley Oil Company in 1928 about two miles south of the Pecos River. Unfortunately, Clayton's well did not strike oil.[11]

Clayton and Waldo also did the surface mapping of an area west and northwest of Fort Stockton that they believed was structurally promising for oil. Clayton did the plane table work and his brother did the geology, creating a map of this area that would years later prove to be an amazingly accurate description of what would become the productive Fort Stockton field.

"Their map was very, very accurate," said Clayton's son Claytie.[12] O. W. was so encouraged by the structure his sons found that he joined with them, J. C., and a friend to acquire 1,120 acres of oil leases in the area, although he entered the project with his usual skepticism. "I am puzzled to understand

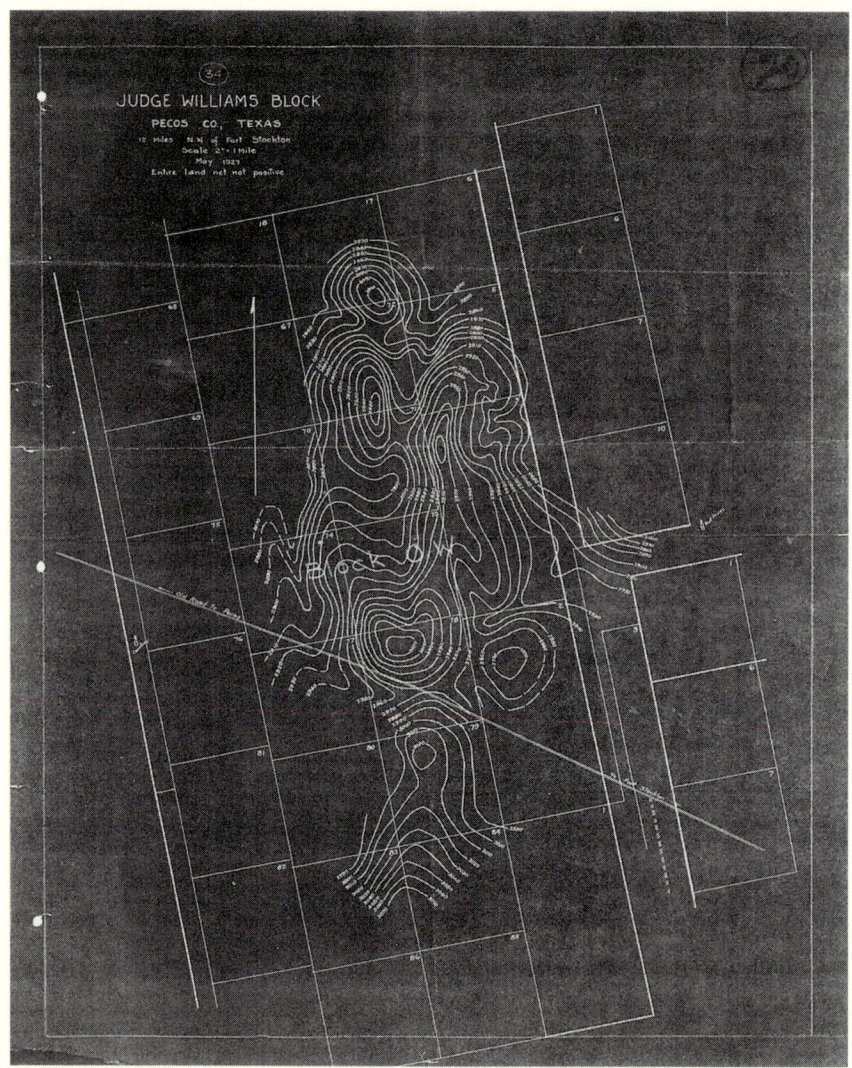

A surface map drawn in 1927 by Clayton and Waldo of an area west and northwest of Fort Stockton that they believed was structurally promising for oil. The brothers were self-taught geologists. Years later, the map proved to be an amazingly accurate description of what would become the Fort Stockton oilfield.

why any oil structure close to town could have been overlooked so long by major companies," O. W. wrote J. C. "Besides this I am discouraged by the fact that only 2 days ago there was a big cut made by the pipe-line Companies in the price of crude oil in the Permian Basin so that the posted price is now only ⅔ of what it was before the cut."[13]

The price cuts resulted from an oil market that was oversupplied due to the heavy exploration that had occurred in the Permian Basin and elsewhere during the 1920s. Major oil purchasers cut back on buying production that involved high transportation costs, including oil from the Permian Basin and the Texas Panhandle. These price cuts, and the economic impact of the Great Depression, would have serious ramifications for the Williams family in the decade of the 1930s.[14] Clayton later drilled a well to about 1,000 feet on a high structure within the family leases northwest of Fort Stockton but ran out of money before striking oil. He sold the family's leases to the Humble Oil and Refining Company. Years later, Humble would drill the discovery well of the prolific Fort Stockton field and strike oil at a location about a mile from Clayton's well. In 1962, the Gomez Ellenburger natural gas field, one of the largest volume natural gas fields in the United States, was discovered at much deeper levels under the Fort Stockton field.[15]

Landreth Gas Fire

Having by now acquired interests in several thousand acres of oil leases near Fort Stockton, O. W. followed Pecos County oil exploration efforts with increased vigilance. He also continued his interest in the technical aspects of oil exploration and production, and described in his letters to J. C. any interesting developments in these areas. One such description detailed how a large fire at a Landreth Production Company natural gas well in the Taylor-Link field east of Fort Stockton was brought under control by Ward A. "Tex" Thornton, the well-known oil and gas firefighter, in the summer of 1929.[16] According to O. W., the driller and his assistant, a tool-dresser, heard a deep rumble before the drill tools exploded up to the surface.[17] The driller yelled at his assistant to run. "But when [the driller] turned for his own flight he saw the tool dresser back 20 feet away set towards the hill, and with such a good start that he was never able to get in talking distance of him in the next half mile. For almost immediately after the gas roared up in the rig, it caught fire, and with the flash of that blaze, the tool dresser's speed was raised beyond that of a jack rabbit, with the driller playing a good second."

The fire burned for nearly a week, and O. W. said the roar was so great that people within a quarter-mile of the well could not hear each other talk. "At night the light brightened up the hill slopes of the Cutoff Mountain— 10 miles away—so that a man might read a newspaper." Thornton was called in to put out the fire, and O. W. learned that the plan was to extinguish the blaze with dynamite. Clayton loaded up his wife, mother, and father into a car and drove out to the well on the night the charge was to be touched off. When they arrived they found about a hundred cars gathered from Fort

Stockton, Rankin, McCamey, Buena Vista, Sheffield, and even San Angelo. But the shoot was postponed and the family returned to Fort Stockton.

O. W. later learned the shoot was set off the next night. "At that time a low platform had been placed about 8 feet from the well, and to it Thornton in his asbestos suit walked from behind one bulwark to another until he came to the platform upon which he placed the nitro-glycerine, to which there seemed to be a cord attached. Then he turned and walked away some 30 or 40 feet—probably to the end of the cord—and was seen to make a rapid, jerking motion of the arm, and then the light went out with the explosion, but only for a moment. Then it came on again as bright as ever, and the performance had to be staged again, which occurred at 3 A.M. the next morning—for the second and successful shot."[18]

Depression, Earthquake Hit Pecos County

It was the summer of 1930 when the impact of low oil prices and the economic depression was being significantly felt in Pecos County. "The oil business is quite slow, and the general situation in Texas is that of hard times," O. W. wrote J . C. in July. "It was due after all [to] the Wall St. craziness, and will require one to three years duration—products way down— wheat under a dollar—Cotton price 60% lower than last fall. . . . I have for once been wise enough to be ready for the storm, but I am carrying Clayton for $12,000+ and paying Waldo—$125 a month—just to keep his family going. Yet we are probably in better shape here than in most Texas counties, owing to the large amount of revenue coming in on [the] Yates oil field & wildcatters."[19] In his letters to J. C., O. W. does not break down the $12,000 owed him by Clayton, but the debt was probably due to a combination of losses in the Pyote Refinery, purchases of oil leases and drilling expenses, and some portion of the construction expense associated with the Crane ice and water operation.

During this period Clayton began spending more and more time at the Crane plant, which by now was his only paying venture and delivered about $5,000 a year. O. W. fretted about Clayton's lack of income, and when J. C. wrote that he was sending a rug for Clayton and Chic's new home, it triggered in O. W. an oft-repeated observation about Clayton's economic situation. "That rug you have ordered for Clayton & Chick [sic] seems to me to be rather extravagant considering Clayton's financial situation, and the appearance of hard times—very hard times—which seem to be settling down on us. Clayton owes me a little over $12,000, [owes] Mrs. Sykes about $5000 and your Mother some $1500. His Ice & Water plant is at its best just now and it is paying him some money—say at a 12 months figure of $5,000 profit. But he lives on that, and you can see that it will be years before that income pays those debts. He must get help from those leases he owns, or sell the Ice & Water plant in order to pay them off. Luckily for him his wife is not extravagant but is disposed to help him out by economy on her part."[20] J. C. eventually shipped three rugs to Clayton from China, sticking O. W. with more than $400 in customs and freight charges. "Do not buy more," O. W. instructed J. C.[21]

As the Pecos County economy was rocked by the depression and falling oil prices, so was the landscape rocked early on the morning of August 16, 1931, by the strongest earthquake in Texas history. It measured about 6.0 on the Richter Scale, was centered near Valentine, about ninety-five miles west of Fort Stockton, and occurred at about 4:45 A.M.[22] O. W. described the experience for J. C. in a letter dated that day, noting that the quake shook Fort Stockton but did little damage. Awake at 4:20 A.M. but not yet out of his bed in the library, O. W. felt the vibration.

> The thought that first came to me was that there was a man or animal under the bed that had turned over suddenly and in turning had rubbed rather hard on the under side of the bed and caused it to vibrate rather strongly. So I leaned out over the side of the bed trying to determine in the shadows what, if anything, could be moving the frame from below. There was nothing under the bed to explain the motion. About that time the first shake ceased and I laid there puzzled over the movement, and wondering if anything could be wrong with my sense of motion—with my nerves—or my feeling. But after about a minute of this lull—there came another shock, more violent and more prolonged, during which I could plainly ascertain that my bed was being shaken in an East and West movement. There was now no doubt in my mind; it could only be an earthquake, even though in my 48 years here we had never had another.[23]

East Texas Oil Boom

With little to show from his leases and having drilled at least two dry wells, Clayton tried to find some steady employment in 1931 beyond what the Crane plant provided, but there was very little work to be had in Pecos County. O. W. described how bad conditions were in a letter to J. C., noting that merchants were selling at a loss just to put a few dollars in circulation. "The rent of rooms and houses—always heretofore too high with us—has dropped to one half of former price. The contracts for big road work have not yet been let. The rainy winter has cut off the demand for alfalfa held over, and the price of cotton is down at the bottom of the sea in that legend-ary 'hole in the bottom of the Sea,' so the farmers are needing wage work before beginning to farm. Nothing yet here to relieve the situation."[24]

Contributing to a decline in oil prices was the discovery of the vast East Texas oil field in the fall of 1930. Covering 134,000 acres in Rusk, Gregg, Upshur, Smith, and Cherokee counties, the field produced a high-gravity oil that was free of sulfur and had excellent gasoline yield. Labor costs were low with out-of-work men flocking to the area, and the average depth of pro-duction in the field was relatively shallow at 3,667 feet, so drilling costs were also low. By June of 1931, the field was producing 350,000 barrels of crude oil daily from more than 700 wells, flooding the market with high-quality crude and subsequently dropping oil prices throughout Texas and the rest of the United States. On July 8, 1931, the posted price paid for high-quality crude by one major oil company was ten cents a barrel. A year earlier, high-gravity crude oil produced in northern Texas had sold for $1.19 a barrel.[25]

Seeing an opportunity in the new oil boom and desperate for a paying proposition, Clayton headed to Kilgore in the spring of 1931 to join his old friend Carl Cromwell in a project to drill for East Texas oil on a twenty-eight-acre lease. With his usual pessimism, and a trace of bitterness at debts not paid, O. W. viewed the venture as ill-fated from the start. "Clayton is off to Kilgore in East Texas (new oil strike) to blow in some money in oil drilling. He will not have a dollar in 5 years. It is no time to drill oil wells at present crude prices & a man cannot sell a good oil well (what would have been good two years ago) at any price. He owes me $12,000—for two years or more—got $12,000 profit last year from his Crane Ice Plant—spent $15,000 in dry holes & never paid me a dollar. And his Crane Plant can hardly pay after 5 years."[26]

To reduce costs, Waldo and his family moved into Clayton's house in Fort Stockton and Chic went to Alpine to live with her mother. "Clayton cannot be at home while handling the well in East Texas & he cannot just now find suitable quarters for Chick [sic] at decent prices near the well, so the present arrangement was made," noted O. W. "A single room costs $40 to $80 a month now while the East Texas boom is on, and Tyler, Longview, and all the small towns near the fields are overflowing—and *tough,* they say."[27] Waldo, unable to get steady work, earned a little money as a director of the Fort Stockton school board and as head of the local Boy Scouts. "He greatly enjoys the work and will be gratefully remembered by many a man in future years, but it is left to us—his kinfolk—to see his own children get a decent introduction to practical life," observed O. W.[28]

The Williams-Cromwell well struck oil in late May, but as O. W. would soon discover, this bit of good news created even more financial problems for

Clayton. Oil production from the well was rated at 29,000 barrels per day. However, by this time the East Texas oil field was engaged in a protracted battle with the Texas Legislature over proration—the effort to limit production to keep oil prices from falling further. The Texas Railroad Commission allowed Clayton's well to produce only about 200 barrels per day. Humble Oil and Refining Company had contracted to buy the oil at thirty-five cents a barrel. With production so limited, O. W. estimated that it would take a year for Clayton to recover the drilling costs.

In addition to the drilling expense, Clayton had also assumed the expense of the lease contract. Cromwell had negotiated the original contract for the leases but was unable to make the necessary payments to the lease broker. So in exchange for Clayton's assuming the lease payments and drilling costs, Cromwell signed over the lease contract to him with a promise to assume half the expense for half their share of the well profits. Unfortunately, Cromwell was broke and invested no money into the project, leaving Clayton with the responsibility of paying off $11,000 to creditors demanding their money. "I hope this lesson will cause [Clayton] to let up on wild—uncalculated—gambling, and to give more attention to cleaning up on his Crane ice plant," O. W. said. "He has always neglected that, although it has kept him going now for over 2 years."[29]

Further complicating matters was a cloud over the title to the land that could invalidate Clayton and Cromwell's leases. At the time Cromwell signed the lease contract, the family leasing the land was being sued by another party who claimed ownership. "Cromwell knew [that] fact, but chanced it on an opinion of a lawyer who, on the affidavit of an interested party, certified that he did not believe the claimant could win," wrote O. W. He harshly summed up the legal problems as "the result of almost criminal negligence" on the part of Clayton and Cromwell, and felt that even in the best of circumstances the case might take two or three years to resolve as it worked its way through the courts. "In the meantime the well must be operated at the expense of Clayton & Cromwell without any income from it, because the pipeline Co. will not pay out the oil values until the suit is finally decided. . . . So you can see that it looks almost hopeless. All because of the lack of exercise of a little hard business sense."[30]

After learning of his son's financial and legal difficulties, O. W. moved quickly to buy time in the hope that events would play out in Clayton's favor. But bad luck continued to plague the venture. Cromwell died in a one-car automobile accident in Pennsylvania on September 27 while traveling back east in search of financing for his share of the well expenses. Many years after Cromwell's death, Clayton told his family that Cromwell was depressed and drinking heavily, and that the driller had taken out a life insurance policy shortly before his death. Clayton said he believed the accident might have been suicide.[31] At the time of Cromwell's death, O. W. said he doubted that Cromwell's estate would be able to contribute any money to pay off Cromwell's obligations to him.

The Williams family's financial situation took another turn for the worse on the morning of October 6, when the First National Bank of Fort Stockton failed to open. According to an article in the *Fort Stockton Pioneer* three days later, runs on banks in El Paso and San Angelo had put tremendous pressure on First National, and depositors, fearing the bank's failure, lined up to withdraw their savings. The family lost nearly $5,000 (equivalent to about $63,000 in 2009) as a result of the bank's failure.[32]

Despite these setbacks, O. W. was able to raise $3,000 and pay off two of Clayton's creditors by selling three 1/64 shares in the East Texas well to others outside the family. He also invested $9,000 of his own money to pay off other debts. There was one lawsuit pending against Clayton for a $2,600 debt, but O. W. noted that it was "held up in Gregg Co. and will be slow to come to judgment." After assuring himself that other creditors would not immediately file suit for payment, he set off on November 2 for Henderson, the county seat of Rusk County, to closely follow the trial being held to determine which of two parties had a legal claim to the land on which Clayton's well was located. If the legal case was decided in Clayton's favor, O. W.'s plan was to sell more 1/64 shares in the project to pay off Clayton's debts.[33]

Bell Reed v. Bradberry et al.

The trial, *Bell Reed v. Bradberry et al.*, dragged on through December in the Rusk County District Court. According to O. W., the family of a man named Bradberry, who had since died, had executed the oil leases as the land owner but had no title of record to the land.[34] The last title of record was held by Bell Reed, and he filed the claim to the land. The Bradberry family argued that Reed had signed over a deed to Bradberry, but the deed had been destroyed in a fire. The family also claimed ownership of the land under the law of adverse possession, on the ground that Bradberry or his heirs had lived on, possessed, and claimed the land against everybody for forty years and paid taxes on it for that period. Reed answered that he had not given a deed to Bradberry but that he had leased the land to him in exchange for thirty-five dollars a year in rent, and that Bradberry had assumed responsibility for paying all taxes until Reed returned to claim the land.

O. W. doubted the validity of Reed's claim, because Reed testified that he had moved out of the county, claimed his thirty-five dollars' annual rent only once, and never seen Bradberry again. Reed said he learned in 1930 that Bradberry had died in 1924. Reed, a poor tenant farmer, said under cross-examination that after leasing the land to Bradberry in 1888, he had moved from rented farm to rented farm year after year, once farming for two years within four miles of Bradberry and the land. O. W. found it implausible that a poor tenant farmer, living some years very close to Bradberry, would not return periodically to claim an annual payment of thirty-five dollars; Reed had come forward only after oil was discovered. O. W. was further skeptical of four witnesses called by Reed's attorney, who testified they had witnessed the lease-signing forty-three years before. "Each man recollected exactly all that Reed testified, but went further and told the exact spot where the leasing agreement was made—25 feet East of the R.R. at a certain crossing."

Testimony from witnesses for the Bradberry family supported the family's contention that Reed had given Bradberry a deed to the land. Bradberry's widow, then eighty-two years old, testified that the deed had been placed in a trunk in the family home, which was destroyed by fire in 1898. Two witnesses testified that they had seen a vendor's lien note dated 1893, signed by Bradberry, and given to Reed for $224 in partial consideration for the deed. A merchant who had bought the note said that Bradberry paid it off in 1899. Other witnesses testified that the general understanding for many years in the neighborhood was that Bradberry owned and claimed the land.

O. W. did not witness the conclusion of the trial, because he was needed at the Scott & White Hospital in Temple, where his sister Sue was having surgery. He learned by telegram of the verdict for the Bradberry family and thus for Clayton. O. W. and Clayton were able to sell Clayton's interest in the well about nine months later for $25,000, enough to pay off Clayton's creditors, although some of the money would be paid in future oil production that, due to proration, would not be realized for some months to come.[35] However, Clayton's son Claytie believes his father's failure to keep his interest in the well dampened his enthusiasm for the oil business. "He stayed somewhat in the oil and gas business, but after that deal it kind of broke his heart because he had a big well."[36]

Living in Crane

While struggling to pull Clayton's financial bacon from the fire of bankruptcy, O. W. continued to worry about Waldo's lack of income and expressed his concern in his 1931 Christmas Day letter to J. C. "He is in hard lines—but I kept him in a salary of $125 a month for 2 years when I had no work for him—and he could not hold his family to that figure, so now when I cannot help—he is in bad shape. He just will not hold his family down to a proper scale of living.[37] O. W. frequently criticized Waldo, his wife, and family for spending beyond their means but never provided any specifics. Clayton, meanwhile, became a father in 1931. From birth Clayton W.

Clayton with son Clayton W. Williams Jr.—forever known as "Claytie"—who was born in 1931. Clayton moved the family from Fort Stockton to Crane the next year to be closer to his ice and water plant.

Williams Jr. was called "Claytie" to distinguish his name from Clayton's nickname, "Clayt."

At the beginning of 1932, Clayton moved Chic, her mother, and Claytie to a small house in Crane so he could work "hands-on" at the ice and water plant and reduce his labor costs. O. W. looked after the new house in Fort Stockton. "Our home in Crane could almost fit into our living room in our home in Fort Stockton," Chic recalled years later. "When the wind blew against the corrugated tin it sounded like rocks being thrown. But we were all well and working and happy. Home is where the heart is. We learned 'togetherness' in an early stage of our marriage."

Although Clayton had acquired some limited experience as a young man while working in a small ice plant in Fort Stockton, the plant in Crane was much larger and more complicated. On many occasions he would have to make the 100-mile round trip from Crane to Odessa to purchase parts for the plant. "Now and then the hour and occasion was right and Mernie [Chic's mother], Claytie and I would go with him," Chic recalled. "This was quite a treat."[38] As the family finances were fragile, Chic did everything she could to economize. Her father, Oscar Henry Harrison Graham, had undergone many boom-and-bust cycles in the land business, and Chic's youth had prepared her for hard times. "One year we would have a six-passenger car and the next year we couldn't pay the grocery bill," she recalled in a 1987 interview with the *Midland Reporter-Telegram*.[39]

During the depression years with Clayton, Chic called on the things she had learned as a youngster when times were tough. "We learned a bit about the 'barter system.' My mama would trade a loaf of her wonderfully delicious light bread to a neighbor, Mrs. Hanks, for a dozen eggs. We ladies sewed up runners in our hose and 'did' our own hair. I'm sure it showed it!" Chic recalled that people who owed Clayton money for water would sometimes pay in merchandise.[40] Clayton also provided some economic support for Chic's family, the Grahams, who were having a difficult time making ends meet. He hired Chic's brother Oscar to work at the Crane plant and provided a roof and a bed for Chic's mother Mernie, while Chic's father and other brother Joe struggled to make a living as farmers and laborers.[41]

As the depression deepened in West Texas in 1932, economic activity in Pecos County ground to a halt, and O. W. took note of the desperation in the lives of Fort Stockton townspeople in a December 1932 letter to J. C.

> Here we have been supported by the ranchmen—the farmers and the railroad—all now 'broke' and without credit. For sometime now the work on the Public Highways through the County has furnished most of the money which has kept the town up but our new Governor Jim Ferguson is putting in all the power he can use on the Courts and the future Legislature to cause a cessation of this work. To stop this work at the beginning of winter will work a desperate hardship on some people of our town. Some are now without food, and almost without credit. The town—already taxed to the constitutional

limit—cannot help them and there is nothing in sight to bring in money before the cattle sales begin next spring and the harvests next summer. Our neighbors of Alpine and Pecos are also in straights [sic].[42]

O. W.'s reference to Jim Ferguson as the incoming governor of Texas is perhaps tongue-in-cheek, because Ferguson's wife, Miriam Amanda Wallace (Ma) Ferguson, was elected to her second term as governor in 1932. James E. (Pa) Ferguson served as governor from 1915–17 and was reelected in 1916 before he was impeached and forbidden to hold any public office in Texas. When Pa Ferguson failed to get his name on the ballot for the 1924 election, Ma Ferguson declared her candidacy and told supporters that if elected she would take the advice of her husband, and that the state was getting two governors for the price of one. She was elected Texas' first woman governor in 1924 but was defeated in the primary election of 1926. She did not seek office in 1928 and was defeated in the Democratic primary for governor in 1930 before being reelected in November 1932. She was a strong fiscal conservative whose policies aimed to reduce state expenditures.[43]

The only ray of sunshine on the economic horizon in Pecos County was increased oil-lease sales activity as speculators acquired ten-year leases at bargain prices from hard-up ranchers. The speculators recognized that once the economy recovered, lease prices would go up. Clayton described the impact of the depression with his typical dry humor. "People were destitute and wishing for the wind to change. But the only change it made is emphasized by this story. An easterner arrived in this country during a windstorm. He vehemently remarked to an old timer: 'Good gracious, does the wind blow this way all the time?' 'No, by George,' replied the old timer. 'It will turn around and blow the other way the rest of the time!'"[44]

The economic hardships continued into 1933. Clayton ran his water and ice plant at Crane and, despite O. W.'s admonitions, continued to speculate in oil exploration and drilling. By now Clayton had acquired a small portable drilling rig and lent it out to oil speculators near Imperial in Crane County, in exchange for a small interest in the oil lease. As usual, O. W. thought this was a bad play, because most wells in that area were small and oil prices were low.[45] He was proven correct a few months later when a small well in which Clayton had an interest came in near the Imperial Reservoir, but the oil could be sold only at 22½ cents a barrel. In addition to providing his drilling rig, Clayton had also covered $800 of the cost of drilling the well, money that, in O. W.'s view, Clayton could ill afford to risk.[46] In April Clayton turned his interest in the well over to O. W. He hoped his father would be able to sell it to recover the money he had provided to extract him from the East Texas well fiasco. O. W. did sell the lease later for $10,000 in cash and $20,000 in future oil production. Clayton later estimated that had he been able to hold his interest in the lease, the property would ultimately have been worth several million dollars.[47]

The hard economic times intensified O. W.'s wariness about Clayton's interest in the oil speculation. "The only good business move that he has made lately was made when he moved over to Crane to see personally after his business at that place," O. W. wrote J. C. in March, 1933. "And it will be some months before that move will do him substantial good."[48] A month later, reflecting further on the Imperial venture, O. W. observed in a letter to J. C., "There is but one Golden Rule in business affairs, and that is the rule of the road 'Safety First.'. . . But it appears to be a most difficult matter to educate Clayton up to that Golden Rule, while Waldo never gets into enough business to call for any application of that Rule."[49]

Waldo, however, had managed by this time to get steady though limited employment on a state highway project. It was located near Sheffield, and he expected it to last until May. He also hoped that when the Sheffield project ended he would be transferred to another highway project running from Leon Springs, just west of San Antonio, to the West Texas town of Balmorhea, and that it would provide him with an additional six months of work. Clayton was optimistic too. He expected increased income from his water plant, because employees with the Gulf Production Company, which had begun a deep test well in Crane County, were bringing their business to Crane.[50]

New Hope for Pecos County Oil

The combined effects of low oil prices and proration, development of the East Texas field and resulting labor shortages elsewhere in Texas, and the general economic depression had brought drilling in Pecos County almost to a standstill in the early 1930s. Even so, O. W. began 1933 with high hopes for the exploration and development of some of the Pecos County oil leases that he and his sons had acquired. Fueling his optimism was the drilling of a deep Pecos County test well by Humble in an effort to find paying quantities of high-gravity crude oil in the Pre-Permian Ordovician formation that had been first discovered by Clayton at the University 1-B well. Humble and Gulf drilled deep test wells in Crockett, Pecos, and Crane counties during this period, including the record-setting 12,786-foot Gulf J. T. McElroy No. 103 at the Upton-Crane county line, but no production was found rivaling that of the University 1-B.[51]

O. W.'s hopes were further raised by the actions of the Sheridan Oil Company of Tulsa, Oklahoma, which in 1932 had acquired several thousand acres of leases west of Fort Stockton near those held by O. W. and his sons. By the beginning of 1933, Sheridan Oil was actively trying to enlist financial help from Humble to drill a test well in the area. Clayton, at the oil company's request, made an oral report to company representatives in June of 1933 on the surface structure west of town that he and Waldo had mapped a few years earlier, followed by a written report with explanatory maps. O. W. also tried to interest the Cordova-Union Oil Corporation of Fort Worth in investing in the project with Sheridan but had no success.[52]

By the summer of 1933, Roosevelt's New Deal programs to improve the nation's economy were beginning to have an effect in West Texas. In July, the citizens of Crane voted to incorporate and acquire a $75,000 loan from the Reconstruction Finance Corporation (RFC) to build an electric plant and a city waterworks. Established during the Hoover administration in 1932 to provide liquidity to the banking system, the RFC's activities were greatly expanded during the New Deal. The proponents of the Crane incorporation advocated acquiring Clayton's water and ice plant for the city. "Now if he could sell his Plant to the City at a fair price that would be the best thing that could happen to him," O. W. noted.[53] O. W., who leased irrigated land to cotton farmers, also saw at firsthand the positive impact of New Deal agricultural subsidies and loans. "The people tell me that I will get more cotton this year from 25 acres than I have been getting from the 40 acres," he wrote J. C. "Quien Sabe? [Who knows?] It looks too good."[54]

The end of the summer was marked by the family's concern over a seriously ill Chic, who was diagnosed with diphtheria and placed in quarantine for two weeks in her house in Fort Stockton, with only a nurse for company. "Clayt and Mernie would hold my precious two-year-old Claytie up to the window for us to see each other," Chic wrote. While Clayton returned to his business in Crane, O. W. watered his son's flowers and lawn, ran errands for the family, and looked after little Claytie. After a month-long convalescence, Chic recovered and Clayton drove the family back to Crane in early September with Claytie riding on his lap.[55] The year ended on an upbeat note when J. C. came home to Fort Stockton from China for a Christmas visit with his family.

With the prospect of selling his waterworks and ice plant to the town of Crane, Clayton returned his family to the new house in Fort Stockton in 1934. Though the economy was still struggling, Clayton and Waldo managed to find reasonably steady employment. Waldo's work on the highway had ended, but he landed a thirty-hours-a-week job as Pecos County engineer for $75 a month. Early in the year Clayton was hired for three months at $110 a month to head up a survey crew for the Coast and Geodetic Survey at the site of the Red Bluff Dam, located about fifty miles northwest of Pecos near the border with New Mexico. In his free time Clayton traveled back to Crane to check on the waterworks and ice plant and visit his family in Fort Stockton. But the major development at this point in Clayton's career was his decision to go into politics and run for Pecos County commissioner.[56]

CHAPTER 10

A Public Figure

P ecos County, like each of the 254 counties in Texas, is governed by a commissioners court composed of four precinct commissioners and a county judge. Commissioners have a variety of responsibilities that include road maintenance, establishing the county's budget and property tax rates, authorizing contracts in the name of the county, establishing the county's employment level, and calling, conducting, and certifying elections. Each commissioner represents a quarter of the county's population and is elected for a two-year term.[1]

Clayton Williams's decision to run for county commissioner was no doubt predicated on his need to hold a steady paying job, but it was also influenced by his belief that he had been treated unfairly by the sitting commission. In 1933 the commission decided to hire someone to evaluate oil properties in the county for tax purposes. At the urging of his father, Clayton applied for the job, but was not hired. He said he decided to run for the commission because, based on his experience in the oil business, he believed that a "clique" on the commission overpaid the person who was hired and was wasting tax dollars.[2]

Clayton announced his candidacy for Precinct 1 county commissioner on February 9, 1934, declaring himself to be "capable and thoroughly qualified to assume any and all duties and responsibilities connected with the office."[3] His opponents in the Democratic primary were Charles Dees, incumbent Fred Cliett, and former commissioner Charles E. Flynt. Clayton won the primary, and, as was the case for most elections in Texas during this period, the winner of the Democratic primary was assured of victory in the general election, which Clayton won by a margin of 258 to 1.[4] His election marked the beginning of a fourteen-year career as Precinct 1 county commissioner, but Clayton's tenure might have been short-lived except for a legal technicality that determined the outcome of the 1936 election. His main opposition in this three-person race was again Fred Cliett, and Clayton won the primary by a seventy-two vote margin. As neither man had a majority of the votes cast, the local Democratic committee declared that a runoff election would be held.

The campaign leading up to the August 22 runoff was reasonably civil, with only vague references by Cliett to wasteful hiring practices by the county for road construction. Clayton responded with a recitation of his skills as an engineer, his honest stewardship of public money, and the successful completion of road projects. However, he also raised the issue of possible election fraud by promising not to buy votes or promise favors to individuals in exchange for votes. This reference reflected Clayton and O. W.'s fear that the Cliett camp was planning to buy votes from ineligible voters, primarily Mexican laborers.[5] "We are not buying [votes], while we have some reason to believe that others are," O. W. wrote J. C. "Certainly much money was spent by them before, and we hear that they have just raised a 'slush fund.' Our political situation is pretty rotten because of those Mexicans—more so than usual."[6]

When the votes were tallied, Clayton fell short of Cliett by six votes. He quickly filed suit in the 83rd District Court, declaring that the primary runoff election should be null and void for a variety of reasons, including voter fraud. After hearing witnesses in the case, Judge C. R. Sutton threw out all of Clayton's allegations about voter fraud but still ruled in his favor. Stutton found that because the county Democratic committee had not formally voted to declare that a runoff election be held, the results of the first primary election should be the only determining factor in who was the Democratic nominee. Cliett appealed to the Eighth Court of Appeals in El Paso, but Sutton's ruling was upheld.[7] Clayton won the general election handily, even though Cliett's supporters bitterly cast sixty-three write-in votes for their candidate.

Clayton was reelected commissioner in 1938 but lost the 1940 race to rancher Frank Hinde in another primary runoff election, this one formally declared by the local Democratic committee. The two men tangled again in a hotly contested 1942 election that again came down to a primary runoff. In this campaign Hinde charged that he found the county commission road fund overdrawn and in the red when he took over the office from Clayton in 1940. Clayton countered that Hinde's figures were wrong and pointed with pride to his accomplishments as commissioner, citing in particular the miles of highway paving and construction that he had supervised. Clayton noted that the county tax rate had not been raised on his watch but had been raised while Hinde was commissioner.[8] Clayton won the runoff by fifty-seven votes and was unopposed in the general election.

Clayton was reelected to the commission in 1944, 1946, and 1948 before deciding not to run in 1950. Years later, in reflecting upon his time as Precinct 1 commissioner, Clayton was most proud of his efforts to improve and upgrade county roads. "One of the first pavements I put in was around the courthouse and around the jail," Clayton recalled. "One of my opponents said it wouldn't last a year and it's still there years later." Janet Pollard recalls that her father was out every day as a commissioner, either supervising the construction of roads or maintaining the roads that had already been completed. "Daddy wore khakis every day. He was always out and working on the roads or in the fields."[9]

Pecos County Oil Exploration Picks Up

By 1934, oil exploration in the Permian Basin was on the upswing, and the major oil companies and some independents were able to increase their exploration activities at deeper levels using rotary drilling.[10] Because these deeper pays were in tight limestone formations, a process called acidizing was used to pump large volumes of hydrochloric acid into the formations to open them up. As a keen observer of new developments in the oilfields, O. W. observed that acidizing would prove particularly effective in Pecos County, where most of the wells produced oil from limestone formations.

O. W. was particularly heartened by the drilling of deep wells by the Gulf Production Company in Crane County, across the river from Pecos County. Like Humble, Gulf was seeking oil pay in the Ordovician zone. One of the wells, Gulf's No. 103 McElroy, was located at the Upton-Crane county line and reached a record depth of 12,786 feet in May 1935. However, it found no commercial oil below the Permian. A second Crane County well, No. 1 Waddell, found production in the Middle Ordovician in February 1936. Both of these wells gave O. W. hope that an Ordovician "high" ran from Crane County southwest through northern Pecos County to the Yates field, and if true, that would increase interest in the leases he held on lands in northern Pecos County. However, no significant production was found along this line. "The few Ellenburger and Simpson discoveries were more significant for the geological data they furnished than for the size of their production," observed oil historian Edgar Wesley Owen.[11]

During the mid-1930s the Williams family made two major plays for oil in Pecos County. The first involved efforts to revive a long-abandoned well near Buena Vista in the northern part of the county. The second was an extensive effort to drill a well on land the family leased west of Fort Stockton and thereby increase the potential value of those leases, which might then be sold.

O. W. called the abandoned well the Perrine well (Perrine was apparently the name of the driller). It was located adjacent to lands where O. W. and Clayton held leases. The drilling crew had hit a large pocket of natural gas at 1,490 feet in the spring of 1929, but at about 1,500 feet the pressure from the gas lifted the drilling tools up 70 feet in the hole, where they were caught on a rock shoulder and hung up while the gas passed around them. There was some showing of oil in addition to gas, but the well was eventually abandoned due to the hung-up tools, which could not be extracted. The discovery of natural gas stimulated the drilling of other wells in the immediate vicinity, but there were no significant findings of either gas or oil around the Perrine well, dashing the hopes O. W. and Clayton had at the time for a good score on the land they leased.[12]

However, during a routine visit to inspect the abandoned well in the early spring of 1934, Clayton found oil flowing out the top of the casing. He and O. W. immediately began negotiations with Hyden Eaton, trustee for the stockholders of the Kansas City, Mexico and Orient Railroad, which owned the land, for a lease to resume drilling at the well. "The main question is of course how and where does the oil come from?" O. W. wrote J. C. "Is it coming below the place where the gas came in? If so why does not the gas also come in? I know little of oil wells, and I have found no expert yet who has offered a plausible explanation. Maybe as we go down we will learn the story, but the general idea seems to be that we will not get it until we get to the bottom and drill by and past the tools left in the hole." While O. W. felt the prospects for a producing well were good, he also conceded that the venture could be the "baseless fabric of a dream."[13]

By early April Clayton had a drilling rig up and was making preparations to begin bailing the oil out of the drill hole to determine the situation. Money was still tight, however, and O. W. encouraged J. C. to invest in the project; he estimated it would cost $3,000 to bring the well in. "The oil is of good grade, and Clayton says that [the well] had over a million feet of gas—so it appears to be a fair gamble."[14] Clayton began removing the oil left in the hole at the end of April, but at that point the story of the Perrine well stops in O. W.'s letters and is never mentioned again, indicating the venture was a failure.

The Well Near Town

In 1935 O. W. and Clayton's hopes for a significant oil play on their leased properties west of Fort Stockton were buoyed yet again as the Sheridan Oil Company of Tulsa renewed its efforts to drill an exploratory well on the Williams family leases. Both Sheridan and the family owned several thousand acres of leases in the area and hoped that a successful well would greatly increase their value. O. W. wrote J. C. in September 1935 that he did not have much hope that a well would strike oil, but he hoped that the mere act of drilling a well would stimulate interest in the surrounding land. "I am tired of paying out $600 to $800 every year on rents, while if a well is drilled I can recoup to some extent by sale of acreage."[15]

Clayton, O. W., and representatives of Sheridan formulated a plan to share the costs of drilling the water well to a depth of 300 feet on acreage owned by O. W. to see if the underground structure looked promising. "The main purpose [of the water well] is to determine something of our chances of success in drilling to 1400 feet," O. W. wrote J. C. "If the formations at the bottom of the water well run high, compared with the showing in other places around the proposed well, then our chances are fair, but if low, then they are not promising."[16] By late September the well had struck water at 316 feet and the drilling was stopped. Based on rock samples retrieved from the well, Clayton and Waldo believed the showing was promising for deeper drilling, but the project stalled as the family attempted to work out the financing.

By January 1936, the Williams family was able to forge a tentative agreement with a driller for a 3,400-foot well on eighty acres O. W. owned west of town. O. W. did not reveal the terms of the contract in his letters, but told J. C. that the family did not have to put up any cash, so the deal no doubt involved giving the driller a combination of acreage near the well and a share of any oil that was discovered, in exchange for the work.[17] However, by April the contract had not been signed and the driller was asking for another delay before setting up the rig.

Fortunately, the Sheridan Oil Company reentered the picture in late April and offered to finance the well with a different driller. Even though the contract was not finalized, O. W. was hopeful. "The well is not yet assured but only probable," he wrote J. C. on April 26. "If successful it means much to the town and I am watching for a chance to invest for you. Other wells are being considered near the town, and failure of this well does not by any means finally settle our chances for a 'boom,' although at present there is no demand for our lots."[18] By late May the contract was finalized, with O. W. contributing the well site and 160 acres of leases to the deal, and drilling began.[19]

O. W. and Clayton followed the drilling with great interest, and Clayton made frequent trips to the drill site to measure the progress. During one visit Clayton got into a fist fight with a tool dresser named Burt Pitts after a disagreement about wages. "The fight followed, Clayton with fists, Pitts with a cane," said O. W. "Clayton had the most scratches when Pitts yelled for help"[20]

Sheridan, Clayton, and O. W. planned to drill to about 1,500 feet, analyze the cuttings, and make a decision about drilling deeper. On Independence Day 1936, the well reached 1,100 feet. "Small signs of oil have been struck several times, but that has happened often in wells that proved dry," O. W. wrote J. C. "What encourages is that the strata continue 'high' as the well goes down, but in the end only the drill can tell the finish. It is a pure wild cat."[21] Unfortunately, the "well near town" as O. W. came to call it, suffered a number of setbacks throughout the remainder of 1936 and into 1937. At 1,515 feet the well hit sulfur water, generally not a good sign, but the decision was made to drill deeper. By late fall they had reached 2,815 feet but the driller lost his drilling tools in the bottom of the hole. Even so, the driller was encouraging when O. W. made a visit to the well on November 14. "He showed me the oil & gas signs coming from just above the tools," O. W. wrote J. C. "His contract binds him to go to [another] 320′ or to marketable oil production, & he hopes to get the production."[22]

The lost tools delayed drilling until February 27, 1937, when the driller finally extracted them. "But he had hardly started to drill yesterday when he lost his tools again and today he is trying to get them out," wrote O. W. to J. C. "Now the showing may not make a well and if so the driller must go on 300′ deeper to the place where we . . . expected to find oil. But it is a little wearing on one's nerves to be halted for 2 months where we had a chance of success."[23] After the driller extracted the second set of tools, drilling continued down to 3,000 feet, where a good showing of oil and natural gas was encountered in late March. But after going five feet farther, the drill struck a heavy flow of water. To shut off the water flow, the hole was plugged at 2,960 feet, a level at which there had been two good showings of oil that had been passed up to go deeper.

The parties decided to "shoot" the hole at that depth with nitroglycerin in the hope of opening up the rock to stimulate the flow of oil. The well flowed oil at the rate of 10 barrels per hour after the shot but soon stopped flowing. Further efforts to stimulate the flow were unsuccessful, and by early summer of 1937 the attempt was abandoned.[24] O. W. still hoped that the family would benefit from the project. "It [the well] at least indicated an oil pool and other wells are to be soon drilled near by and if they prove the pool I will get my $10,000 back [O. W.'s investment in the project and in leases near town] and Clayton will get some money, as he too holds leases in the neighborhood."[25]

The Good and the Sad

Despite the failure of these two ventures, the financial fortunes of the Williams family began to improve during the mid-1930s. In February of 1935 O. W. received his first check from his oil interests in the East Texas well in Rusk County, where he had covered Clayton's debts in the venture. "I doubt if I get enough out of the full oil runs to pay me in full for what I put in (over $20,000), but it will help greatly," O. W. noted with his usual pessimism, "and the experience should be a valuable lesson in caution to Clayton."[26] As economic conditions in the county improved, O. W. also began to recover some of the money he had loaned to others during the depression. Due to the increased oil exploration activity in northern Pecos County, he was able to negotiate some oil leases on Williams family lands located near the new activity. "It becomes more evident every day that Pecos County will have more oil development this year than in any previous year so far as the county at large is considered," he wrote J. C. in 1936. "So far the Yates field has represented $99/100$ of our oil activity, but now the other areas—south, west and north—are coming into the play. This is specially the case in the north and along the Pecos River."[27]

Clayton also began to experience a reversal of fortune during this period. On September 21, 1935, Clayton and Chic's second child, daughter Janet, was born in Fort Stockton. And after long and protracted negotiations with the Town of Crane, Clayton sold the water plant portion of his water and ice plant to the city for $12,000 (equivalent to $188,000 in 2009) in the spring of 1936 and continued to operate the ice production portion. He also was heartened that his father was receiving a steady stream of oil payments from the East Texas well, and by the summer of 1936 Clayton's considerable debt to O. W. had been reduced by $9,000.[28]

Matters also began looking up for Waldo, who in the summer of 1935 secured a $170 a month job as an engineer to work on the Red Bluff Dam near the Texas-New Mexico border in Loving County, Texas. The dam, located on the Pecos River, was funded by $2 million in bonds sold to the Reconstruction Finance Corporation, and was designed to provide water both for irrigation and production of hydroelectric power. It was completed in 1936.[29]

Despite improvements in his financial condition, O. W.'s hopes for the nation's moral condition suffered a setback in 1935. Texas voters, riding the wave of the national approval of the Twenty-First Amendment overturning prohibition, ratified a repeal of a prohibition amendment to the state constitution that had been passed in 1919. Calling the repeal "a very distinct step backward," O. W. saw it as "about the worst result of our Depression, as it is claimed that taxation from liquor will pay $5,000,000 and so reduce taxes on other property. That is a good line or argument for licensing [the] sale of opium &c." A month later, O. W. advised the still-unmarried J. C. to

*Clayton and Chic's daughter Janet with grandfather O. W. Janet was born in 1935.
She spent much of her time as a youngster with her grandfather, whom
she nicknamed "Judgie."*

"Take a serious minded young woman for a wife, who has only moderate social ambitions and an aversion to liquors."[30]

Tragedy struck the family in the summer of 1936 when Sallie, O. W.'s wife of nearly 55 years, died of a heart attack on August 25. Hospitalized in San Angelo for three days with a throat obstruction a week before her death, Sallie returned to her home with a throat laceration but was, in O. W.'s words, "apparently strong." In his typical attention to detail and to leaving a historical record of her death, O. W. described her last days to J. C. in a letter dated August 30:

> Saturday was the day of the Primary Run-Off [for Clayton's County Commission seat] at which I was present all day. I left hom [*sic*] understanding that she would not come out to vote, so I was surprised when she came to the polls to vote. Up to that time and until the next Tuesday she appeared to be about as when she came back except that she was not eating sufficiently, which had been attributed to the condition of her throat. But that day the doctor told me there might be a complication due to some adrenal difficulty relating to the condition of her kidneys. This was the condition Wednesday night about 9:30 P.M. after I had gone to bed and Kathryn [O. W.'s daughter] and Chick [*sic*] were sitting up with her, when she complained of feeling much heat, and asked them to help her up to a sitting position. They did so, and then in a few moments she collapsed to the bed, and without a word passed out so quietly and easily that it was some time before they realized that she was dead. It was a great shock to me. I had always thought I would go first.
>
> Of the 66 years I have spent working for myself—hard times most of them—she has spent nearly fifty-five of them with me, loyally and cheerfully, sunshine and storm, as you must remember; although I took her from a good home in Dallas, Texas, into a life of hardship and exposure in the Frontier for the remainder of her life. When you think of your mother hereafter always remember that story. If I did not think there is a reward for her hereafter I should blame myself for her hardships in this life, and feel that there is no such thing as Eternal Justice. God's mercy must surely rest on her.[31]

Sadly, articles were published side-by-side in *The Fort Stockton Pioneer* on August 28 detailing the election results and the death of Clayton's mother.

O. W. buried his wife in East Hill Cemetery on a small hill south of Fort Stockton. The funeral procession stretched back to town, a distance of one and one-half miles. At the head of her grave he had a granite marker placed on which were inscribed her full name, parents' names, O. W.'s name and date of their marriage, and the dates of her birth and death. A few weeks after her funeral, J. C., who was still in China and could not attend the funeral, expressed his regrets at one task left undone. "From a small photo Ermine [his sister] sent me last year I had a Chinese artist make a colored portrait of Mother, dressed in the gorgeous gown of an Empress Dowager. My pleasure was in anticipating the surprise Mother would show in seeing herself so decked out. I'm so sorry I didn't send it right away."[32]

The year would not end without a second tragedy. After a long and difficult fight against lung cancer, including experimental radiation treatments in California, Waldo's wife of twenty-three years, Olive, died on October 12. "Olive passed into a coma yesterday and died at 7:15 P.M. without regaining consciousness," O. W. wrote J. C. the day after her death. Although he had criticized Olive for overspending her family's budget, O. W. now composed a touching tribute. "She leaves behind 3 children who, we hope and believe, will be better monuments to her goodness and self-sacrifice than any stone which we may set up to her memory. She was a loyal and devoted wife and mother."

Life Goes On

After the death of their mother and Olive, Clayton and Ermine devoted considerable time and effort to helping their eighty-three-year-old father deal with the two losses. "I do not know how I would have fared without her help and that of Clayton," O. W. wrote J. C. in November. "When you consider such children can you wonder that I am anxious for you to marry? Think of my situation had I been a 'lone wolf.'"[33]

But the emotional strain of the deaths, coupled with the stress of earning a living at the tail end of the depression, was putting additional pressure on the family, and particularly on Clayton and Waldo.

Waldo's job at the Red Bluff Dam ended just days before his wife's death, though he did find temporary employment in late November with the State Water Control Board. Clayton, meanwhile, was reelected to the county commission in November after contesting the results of the October primary election. During this time he was also involved in a lawsuit over the amount of his federal income taxes and was in the middle of another lawsuit brought by one of his employees at the Crane ice plant, who had been injured on the job. Chic was also suffering from health problems, resulting from the birth of Janet, and was scheduled for surgery in Temple after the first of the year.[34]

In a letter to J. C., Ermine noted the effects that the stress was having on her brothers. "The boys, Waldo and Clayton, are at cross purposes and show effects of long strain and are now suffering with nerves and are quite hard to get along with—particularly Clayton. You see if not carefully managed these two can split our family. It is an acute crisis—and I am hoping for the best natural result of long strain and shocks. They have broken under it. Clayton has had too much election, oil well trouble, and Waldo has had too much criticism [and] too little relaxing over period of years. Now we hope to overcome this little internal trouble, but I am wildly nervous, and unfit to handle the case."[35]

As the immediate shock of his wife's death began to recede, O. W. contemplated a life without his constant companion of more than five decades. "When [Ermine] leaves I will live by myself, eating with Clayton, or down town, and having my books for companionship as I do not go out after night. I appear to be in good health, but will take a physical examination now and then while I am trying to put my financial affairs in the most convenient shape possible to me for handling after I pass away."[36] He also began making plans for J. C.'s temporary return from China for a vacation of several months in the fall of 1937. In previous letters to his son, O. W. had discussed the possibility of their visiting "the seats of our family" together in Kentucky, Illinois, Virginia, and West Virginia, on a trip similar to the one he had made in 1923. He was now seriously planning the excursion, and he was also considering returning to China with J. C. in the spring of 1938 for a three-month visit "as my last gratification of the wanderlust. But it is a question, whether or not I can get away from the care of my affairs that long."[37]

Indeed, business had improved for O. W., who estimated his net worth at $75,000 at the end of 1936 (equivalent to more that $1.1 million in 2009). By the summer of 1937 O. W. had leased all but 1,000 acres of his land to oil developers and had earned more than $15,000 in payments from the leased lands. Oil payments from Clayton's East Texas well had by now paid off the $20,000 loan he owed O. W., and the remaining payments went directly to Clayton. To O. W., the future for oil development in the county looked more promising than ever, and he noted that much of the land west of Fort Stockton had been leased at prices from $2 to $7.50 an acre. "It is the warmest time on that line that I have ever seen about Ft. Stockton. The Grandfalls country [just north of the Pecos River in Ward County] began to boom some 3 years ago, and now there is a forest of derricks for over 10 miles from there to the Northwest. Monahans [in Ward County] is a big town and 5 or 6 new towns have come into being. Midland is a city with 10 story buildings. All coming from the development of oil in the Permian Basin."[38]

Clayton, meanwhile, had turned his energies to tending the ice plant in Crane, resuming his duties as county commissioner, and worrying about Chic, who after surgery spent nearly three weeks in the Temple hospital with Ermine at her bedside. Hard times continued for Waldo, who lost his job with the State Water Control Board and was found alone at his house in late January delirious with a high fever. "He . . . is naturally rather moody," O. W. wrote J. C. "You should write him up some, although, as you know, he rarely writes to anyone and even less to his kinfolk. A kind of 'lone wolf.'"[39] After recovering from his illness, Waldo traveled to Mexico to find work but was unsuccessful and returned to Fort Stockton, where O. W. hired him to survey some of the family's land holdings. O. W. and Ermine attempted to secure Waldo work in the new national park being planned for the Big Bend region of Texas, but that effort failed. Waldo also expressed a brief interest in taking a course in geology at the University of Texas at Austin to qualify him for employment in the Pecos County oil fields, but nothing came of this idea either.[40]

Ranching Adventures

As the decade came to an end, Clayton was reelected to his third term on the county commission. He continued to operate the ice plant at Crane, looked for opportunities to buy and sell oil leases and drill wells, and with encouragement from Chic, began to turn his eye toward ranching. He began slowly at first, buying some cattle that were being tended by his brother-in-law Oscar in the small town of Quemado, just north of Eagle Pass. "I don't know how I decided on what type of cattle I wanted," Clayton said years later. "I don't know how we happened to buy them but Chic always wanted me to go into the ranching business."[41]

In 1938 and 1939, Clayton and J. C. bought twelve sections (640 acres compose a section) of dry ranchland west of Fort Stockton near Belding and formed a partnership called Wilbroco (Williams Brothers Company). Financed with bank loans that eventually totaled $87,000, they bought both cattle and sheep, but soon kept only cattle on the land. "We had sheep to start with but the town dogs would run out there and give us a fit," Clayton recalled later. "They would just leave dead sheep strung along the fence. So we quit sheep and went to cattle as the cattle were not subject to being cut down by the dogs, but our ranching was not much of a success. I think we managed to pay off our loan but that was about all."[42]

While not a financial success in Clayton's eyes, the ranch provided many an adventure for his young children. Maintaining it required Clayton to

High stacks of alfalfa bales at the Wilbroco ranch made a fun roost for the Williams family children and their friends in the early 1940s. At the bottom of the stack, from left, are Robin Johnson and Janet Williams Pollard. Sitting on top of the stack are Carl (Buddy) Butz, Walter Scudday, James Scudday, and Claytie.

purchase several cow horses from a family friend, including a few that had never been ridden. Shortly after the horses arrived at the ranch, Clayton's son Claytie roped one of the "green" horses and was able to ride it. Janet recalls, "Claytie mentioned to our dad that he had ridden one of these half-broncs. My dad said he wanted to go to the ranch and watch Claytie to be sure he was safe. Naturally, I tagged along. Daddy had me get up on the fence while Claytie roped and saddled the horse. When he put his foot in the stirrup, the horse reared up and started pawing the air. They tried to tie him to a post but the horse continued to rear up and paw closer to both of them. My brother was the closest to the horse and its hoofs were getting very near to his head. Finally Claytie screamed, 'Damn it, Daddy, if you can't run, get out of the way of someone who can.' I got a big kick out of that."[43] Though he knew how to ride, Clayton was not as fond of riding as his son. "He never thought of himself as a cowboy," says Janet, "but he loved the land."

Clayton experimented with different measures to capture water from the infrequent rains and tried growing several different types of grass to see what would work best on the dry land. Once, according to Janet, J. C. sent his brother grass seed from Australia, and the children and Chic were frequently enlisted to help spread new seed. "Daddy's heart may not have been in ranching, but he was always experimenting. He and J. C. were always trying to upgrade what they bought. Whenever it rained we would sow grass seed."[44] A snowfall in 1941 provided what Clayton thought would be an ideal environment to broadcast some grass seed. After dropping off Claytie and Janet at the town movie theatre, Clayton and Chic drove out to the ranch, but their car became stuck. "First, I got in the car and Clayton pushed," Chic recalled. "Then he said we should push in the accelerator as far as it will go and both push. That didn't work. Then he jacked up the back end of the car with a jack. He asked me to get on one side of the car in the back and he got on the other and I said, 'Clayt, this thing might fall on me.' He said it wouldn't fall on me, so I minded him and sure enough it fell on my head. If it had hit me anywhere else it would have killed me."

Clayton walked to a neighboring farm for help, and eventually the car was on the road again. "We had been stuck in the mud all afternoon with our children at the movies," Chic said. "They had seen the show over and over and over."[45]

It was at the neighbor's farm that Clayton learned about the attack on Pearl Harbor.

World War II

With America's entry into World War II, Clayton was offered a commission as a captain in the U.S. Army Reserve, but he refused. "I had too much responsibility then to go back in," he said, pointing out that he now had a wife and two children, an elderly father, and Chic's mother, who lived with the family. "However, if they had offered to commission me as a major, it would have been hard to turn down. We didn't have much money back then." J. C. served during the war as a colonel and chief intelligence office for Maj. Gen. Claire Lee Chennault's Flying Tigers, three squadrons of American aviators who after Pearl Harbor fought the Japanese in southeast Asia and western China. In his memoirs, General Chennault was critical of many of his staff but included J. C. in a list of eight officers who he said were "exceptions" to the rule of incompetence, citing J. C.'s "practical background in the Orient."[46]

When Clayton was reelected to the county commission in 1942 and 1944, the position provided much-needed income; oil exploration activity, except by a few of the largest companies, slowed drastically in the Permian Basin before and during the war. This was due to a loss of foreign markets, the institution of price controls, and the inability to obtain materials and equipment because of federal prioritization for the war effort. "Wells are not being drilled as we expected," O. W. wrote in 1942, "due they say to [a] shortage of iron to make casing & other well [necessities]." According to Permian Basin oil historian Myres, the number of wells drilled in West Texas in 1942 decreased by 44 percent from the previous year.[47]

In 1942 the Central Flying Training Command constructed Gibbs Field about a mile northwest of Fort Stockton to provide pilot training for the U.S. Army Air Force. It was one of a number of airfield training facilities built in Texas during the war. Both Clayton and Waldo provided surveying and engineering services for the water and sewer lines to the airfield. Clayton served as county chairman of the United Service Organization (USO), and Clayton and Chic housed four civilian flight instructors at their home in Fort Stockton while the airfield barracks were being built. "There must be over 200 cadets here in training, and on Sundays our park is full of uniforms," O. W. wrote in May 1942. Two months later he remarked, "We have the daily hum of airplanes now."[48]

The ranch continued to occupy much of Clayton's time and energy during the war. In addition to his experiments with different grasses and various water management techniques, he became one of the first ranchers in Pecos County to breed Brangus cattle. He tried his hand at raising goats, but this effort benefited young Claytie more than his father. Clayton acquired the goats in the fall of 1942 with an $18,000 loan from the Marfa Production Company, a considerable sum for the rancher. As luck would have it, the weather unexpectedly turned cold and rainy after the goats were sheared. All of the sheared goats died, but the family managed to rescue some 250 kids that still had their hair. Claytie raised the goats and sold them for $1,500, but the loss was a considerable blow to the family resources.[49]

Death of O. W. Williams

O.W. celebrated his ninetieth birthday on St. Patrick's Day of 1942. Even at this advanced age, Clayton later recalled, "his reflexes, health and mind had been in excellent preservation due to sufficient exercise, temperance in all things and a course of doing what he could about all troubles, and then not worrying about them otherwise."[50] But on the morning of August 2, while watering Clayton's lawn, O. W. stumbled and fell into an empty lily pond on the property, striking his head on a concrete wall and suffering a concussion and some bleeding in the brain. Although he recovered physically, the accident marked the beginning of a long decline in his mental condition.

"Ermine has been here for over a week helping me out," O. W. wrote J. C. in September. "I am rather weak and can do but little in my 90th year."[51] Clayton said that his father became addled and started wandering but still went to his office daily and, with his son's assistance, tended to his business affairs. "He was reported to have walked down to a neighbor's, remarked as to the neighbor's needs and to his own good financial condition, and offered to give the neighbor considerable of his property," Clayton recalled.

Waldo, now suffering from stomach cancer that would eventually kill him, moved into O. W.'s house with his second wife Gaynell, where she cared for them both. No doubt realizing that his life was drawing to a close, O. W. began the process of distributing his holdings among his children. In the last of his letters to J. C. in 1943, O. W.'s handwriting became weak and almost illegible and the letters, shorter and shorter. As Waldo's condition worsened, he and Gaynell moved out of the house so she could devote full attention to her husband. But O. W. refused to move into Clayton's home, even though his physical condition began to match his mental decline. He suffered severe bouts of pneumonia on several occasions but managed to survive into his ninety-fourth year. During the latter stages of his illness, when O. W. would become restless and confused, Ermine, Kathryn, and Clayton would stand just outside the door to his hospital room and sing old hymns in harmony. It seemed to quiet him.[52]

When Waldo died on August 10, 1946, at sixty-three, O. W. was unable to attend the funeral. A few weeks later, he fell and broke his hip. Confined to his bed, he contracted pneumonia and died on October 29. He was buried next to his wife in East Hill Cemetery. The inscription on his gravestone describes him as "Lawyer, Surveyor, Historian, Naturalist and Writer." Years after his death, a former law partner succinctly summed up the judge's life:

Born on St. Patrick's Day of 1853, he died in the middle of the
1940s leaving to his descendants a rich heritage, not only of
this world's goods, but also of character and culture. He had
built up and made worthwhile use of one of the most complete professional and home libraries in that part of the country. In the midst of a very busy life of gainful employment he
had found time to write a number of pamphlets, many of them
printed, about the great West, its people, its traditions and its
wildlife, past and present, its soil, everything about it. One day
while I was with him in the practice, an Episcopal minister,
pastor of a church in Marfa, Texas, dropped in the office to
meet and chat with a man of whom he had heard so much of
interest. The judge, observing that, 'It is not often that one of
your cloth turns up in our midst,' proceeded to tell the visitor
as much about the Episcopal Church as the minister himself
knew.

Judge Williams was, on all important subjects, the best informed man I have ever known.[53]

Comanche Springs

During O. W.'s final years, Clayton and J. C. sold off many of their land
holdings west of town, but after the war ended they continued to grow
cotton and alfalfa on land irrigated by pumping underground water. The
presence of these wells would, in the summer of 1954, create two of the defin-
ing moments of Clayton's life. In June of that year, the Texas Eighth Court
of Appeals upheld the right of Clayton and other farmers to pump unlimited
quantities of water for irrigation.[54] Two months later, this court victory and
the link between the pumping and the drying up of Comanche Springs—
the Fort Stockton landmark—cost Clayton his race for Pecos County judge,
the same position once held by his father, and marked the end of fourteen
years of public service.

As noted earlier, Comanche Springs, located in the southeastern part
of Fort Stockton, had served as a water source and campground for Indian
tribes on the Comanche Trail to Mexico. In 1849 a reconnaissance party
led by U.S. Army Capt. William Henry Chase Whiting had reached the
springs, and in 1859 Fort Stockton was established by the U.S. Army. The
town of Fort Stockton had grown up around Comanche Springs, and the
springs provided irrigation water to some hundred farmers east and north of
town who formed the Pecos County Water Control and Improvement District Number One. Water was diverted into a concrete irrigation canal and

An irrigation ditch distributing water across land that Clayton and his brother J. C. used to first raise sheep and then cattle, pima cotton, and alfalfa. Beginning in the late 1930s, Clayton and his brother J. C. formed Wilbroco (Williams Brothers Company) and bought and leased thousands of acres. To irrigate, they pumped from an underground aquifer.

transported by gravity from the springs to the farms, where it was equitably divided among the farmers who grew fruit crops, cotton, and alfalfa.

The springs also created a popular swimming hole for the town. It was said that the force of the water flow emanating from the largest spring, Comanche Chief, was so strong that a boy could lie on top of it and be supported.[55] Comanche Springs provided the impetus for the creation of the Fort Stockton Water Carnival in 1936 to help celebrate the Texas Centennial Celebration. The carnival became an annual event, and a county bathhouse, swimming area, and pavilion were constructed at the springs in 1938. In addition to a parade, the carnival included swimming and diving events, music, dances, horse racing, and a bathing beauty review. The 1947 carnival was dedicated to O. W., who had died the previous year. At the 1949 carnival, some 150 volunteers produced "The Cavalcade of Comanche Springs," which dramatized the history of Fort Stockton. The carnival continues to be held in July.[56]

Clayton and J. C. were among nineteen land owners who owned a total of about 37,000 acres of land west of town near the community of Belding.

Of these acres 6,250 were being irrigated.[57] There Clayton and J. C. farmed alfalfa and pima cotton, a strong, high-grade cotton that was developed from selected Egyptian cottons by the United States Department of Agriculture in 1908.[58] In the late 1940s, the Williams brothers, along with other farmers in the area, began drilling water wells outfitted with powerful pumps to irrigate their land. They had purchased the land from three eastern investors who discovered the considerable water resources through test drilling but were unwilling to invest in the expense of drilling wells. By 1951, Clayton, J. C., and others in the Belding area had drilled a total of twenty-one wells and were planning more.[59]

The Edwards-Trinity aquifer supplies most of the water needs of Pecos County, except in the northern part of the county, where the aquifer is absent. The Rustler aquifer, located at a greater depth than Edwards-Trinity, also supplies water for irrigation in western Pecos County.[60] By 1951, officials of the water control and improvement district had begun to correlate the declining flow of water from Comanche Springs with the spring pumping on the irrigated lands west of Fort Stockton. The farms to the east and north of

An early pumping unit used on the Wilbroco ranch to draw water for the farm. The Pecos County Water Control and Improvement District sued Clayton, J. C., and other farmers in 1954, claiming that the pumping caused Comanche Springs to stop flowing. The courts upheld Clayton's right to pump unlimited quantities of water from under his land, citing a 1904 Texas law based on the English common law known as the "rule of capture."

town were now not getting enough water for their irrigation. That year the district began to try various measures to increase the flow of the springs to them, including building and operating pumps at the mouth of the springs, but all efforts failed. The decreased water flow and lower water levels in Comanche Springs also forced the town of Fort Stockton to replace the familiar swimming hole. In 1953 the town constructed a concrete municipal swimming pool served by municipal water. The springs stopped flowing completely in 1961, with only brief periods of flow since. A report that year by the Texas Board of Water Engineers correlated decreased flow from the springs with irrigation pumpage.[61]

"I had no idea that [the wells] would dry up the creek," Clayton remarked late in his life. "I thought that it might weaken a little bit, but I had no idea. When the farmers got after me about it, I told them, 'Well, you would have done the same thing. You would have capitalized on what you had.' And they admitted they would have."[62]

Attorneys representing the county water control and improvement district filed suit in the 83rd District Court in December 1951 to enjoin Clayton and his fellow farmers from taking water from the underground aquifer to irrigate their land to the detriment of the farmers who relied on Comanche Springs for irrigation. But the court held that under a 1904 Texas law, based on English common law known as the "rule of capture," a landowner had the right to pump unlimited quantities of water from under his land, regardless of the impact on others.[63] The water control and improvement district appealed the ruling to the Eighth Court of Appeals at El Paso, and oral arguments were held on April 1, 1954. The appeals court issued its ruling in July, upholding the ruling of the district court.[64] The decision was unsuccessfully appealed to the Texas Supreme Court.

Claytie recalled in his 2007 biography that his father was distressed over the suit and the hard feelings it created in Fort Stockton. "It was a bitter suit with many hard feelings and split our little town for years. My dad was never proud of drying up the springs, if he did, but he did believe that a landowner had the right to the oil, gas, water, and other minerals under his land, the so-called law of capture."[65] Grandson Clay Pollard notes that Texas was suffering a terrible drought when the springs dried up. "I think the drought was a big factor, but no one ever mentions it. You can sue people but you can't sue Mother Nature. I always felt that by not mentioning the drought my grandfather was unfairly criticized."

From 1949 to 1957, the drought in Texas, as well as in parts of the Midwest, was one of the worst in recorded history. It was a significant event in Texas history and formed the backdrop for Elmer Kelton's novel *The Time It Never Rained*. According to the National Climatic Data Center (NCDC),

Texas rainfall dropped by 40 percent between 1949 and 1951, and by 1953, 75 percent of Texas recorded below-normal rainfall. While the area supplying the Edwards-Trinity aquifer is much larger than Pecos County, data from the center show that in 1949 the county's rainfall was about one-half inch below normal. Data for 1950 and 1951 are incomplete, but for the last five months of 1951, Pecos County rainfall was more than five inches below normal. The worst rainfall year for the county during the drought period was 1956, when the rainfall deficit was nearly nine inches. The impact of these deficits was significant for a region that received an average of 13.65 inches of rainfall a year between 1910 and 1959.[66]

Clayton's grandson Scott Pollard, a professional petroleum geologist, says, "If you look at an aquifer as a bucket, then a spring would be a leak near the top or the waterline of the bucket. The people who wanted to protect the springs were saying it was okay to take water spilling out on the surface to irrigate with, but it was not okay to capture water underneath your land. While the sentiments are understandable, most water rights historically have been based on upstream rights over downstream, the right to capture, and then historical water usage."[67]

Janet Pollard believes that her father fought so hard to protect his rights to the water because of what he had witnessed as a child when O. W. had to quit farming because of the upstream diversion of water on the Pecos River. Clayton carried his belief in property-owner rights to his grave. Clay Pollard recalls that one summer while he was visiting his grandparents in Fort Stockton, a representative of a state agency came to the front door and requested permission to go onto Clayton's land to look for evidence of the endangered pupfish. "My grandfather wouldn't give him permission. He said, 'The pupfish is of no value, but if they find it on my land they will take away my rights over my land so they can research and control the pupfish. It is not a good deal to let the government on your land.'"[68]

Many years after Clayton and J. C. sold the irrigated land, Claytie bought or leased several thousand acres in the Belding area that included 2,700 acres once owned by his father and J. C.[69] He noted in his biography that due to the creation of groundwater districts and to modifications in Texas groundwater law made by the state legislature, he is limited in the amount of water that he can pump in a year from the aquifer for his alfalfa farming operations there.[70] Recent newspaper reports have followed an effort by Claytie to have water from his land near Fort Stockton pumped, treated, and moved to Midland in a pipeline. The Pecos County Commissioners Court and Fort Stockton city officials opposed the proposal. In response to criticism of the plan, Claytie wrote in the *Midland Reporter-Telegram* that his company, Fort Stockton Holdings LP, submitted an application to the Middle Pecos

Groundwater Conservation District asking only to use the same amount of water that he uses for agriculture and "to transfer it to municipalities and industries where there is a great need for water. I have made it clear I am not seeking to drill any new wells or withdraw more water than is already permitted to me for agricultural uses."[71]

Comanche Springs Backlash

Many of the farmers whose land depended upon the water from Comanche Springs sold their farms and moved away. But according to George Baker, who published the *Fort Stockton Pioneer* during this period, before the farmers left they organized to oppose Clayton when he ran for county judge in 1954. Baker said in the *Texas Observer* that the water controversy "was far and away the more decisive issue."[72] A second factor contributing to the election loss were two political blunders made by the Williams's camp in the campaign.

Those blunders stemmed from a legal action taken a year earlier by Pecos County District Attorney Travers Crompton against Clayton's opponent, incumbent county judge Paul Counts. Crompton had convinced a grand jury to indict Counts on charges that he had misappropriated county funds by accepting a salary for his duties as ex-officio superintendent of public education for the county. Counts contended that the statutes governing the salary issue were unclear, and he filed a civil suit to resolve the matter when the indictments were issued. After forty-five minutes of deliberation, a jury cleared Counts of all charges.[73]

Clayton announced his candidacy for county judge with a pledge of "Good Government, Good Churches, Good Schools, Good Roads, Good Business," but his campaign became decidedly more pointed after Counts won the primary by fifty-nine votes, necessitating a runoff. In a large political advertisement published in the *Fort Stockton Pioneer* nine days before the runoff election, Clayton addressed a number of pointed questions to his opponent. Unfortunately for Clayton's campaign, two of the questions were based on errors of fact. The first error was in a question which implied that the court had not officially dropped indictments against Counts—Judge John C. Epperson, the trial judge in Counts's case, had made an administrative error that Clayton did not detect. The second error was a statement that Counts had never filed his civil suit to clear up the statutes concerning the salary issue. However, Counts had filed the suit on July 30, 1954.[74]

Both these errors were publicly brought to light two days before the election in a signed page-one article by Baker in the *Fort Stockton Pioneer*. The article noted that Clayton's advertisement "contained or implied errors of fact which the *Pioneer* wishes to correct in fairness to all parties concerned."

In that same edition, Clayton reran the original advertisements with changes that eliminated the errors, but the damage was done. Counts won the runoff election by a margin of 174 votes. Chic would later comment to the *Texas Observer* that "It was the worst defeat [Clayton] ever had."[75]

Clayton took the loss with equanimity. "I did not have any business running for any county office and the voters just put me on notice," Clayton wrote his sister Kathryn. "I should have known it anyway. The results suit me all right as I have more pleasant things to do than to run for political office."[76] Son Claytie remembers that after the election his father said, "'I'm all right. Give me three or four days to put my head under my wing and then we will go forward,' and he did. He was a fair politician, a hell of a lot better than I ever turned out to be."[77]

Eventually, the memory of the Comanche Springs lawsuit and its political fallout faded into the past. Twenty-one years after losing the election, Clayton was named Outstanding Citizen of Fort Stockton, and a few years later he and Chic rode as grand marshals in the Fort Stockton Water Carnival parade. Clayton's daughter Janet and Counts's son Walter are today good friends. And while the loss of the judge's race marked the end of his political career, it gave Clayton the opportunity to fully engage with what would become one of his proudest accomplishments—writing the story of the land and people of his beloved West Texas.

CHAPTER 11

A Love of History

By 1955, Clayton and J. C. had sold the stock from their ranching operation west of Fort Stockton and leased out the remaining farmland. Clayton, according to his children, never loved farming and ranching but recognized the value of the water located under his land and was able to sell at a good profit. Claytie notes that his father's decision to sell the ranch and farm in Fort Stockton was probably the "greatest favor he ever did" for him, because Claytie says he knew then that he "wasn't going to be a cowboy or a farmer anymore and . . . got into the oil business."[1]

At age sixty, with money in the bank and both of his children grown and married, Clayton looked for something else to do. Ironically, his first impulse was to invest in another ranch. J. C. had seen a *Wall Street Journal* advertisement about a ranch for sale near La Grande in northeast Oregon, located between the Wallowa and Blue Mountains. It was owned by the family of former Oregon Governor Walter M. Pierce, who had died the previous year. J. C. made a down payment on the ranch before Clayton and Claytie (now a corporal in the U.S. Army) inspected the property to determine its potential as an investment.

"I had a pretty good nose for business because my dad had let me run the farm and ranch," Claytie recalls. "So I got up there and said this is a good buy at $10 an acre and if you lease it out you'll get a pretty good rate, so I think you should go ahead and buy it and pay for it."[2] Clayton went in with J. C. and purchased half of the property, which they leased out for cattle ranching.

"It was a small ranch about six miles long and two miles wide with two streams running through it," Clayton recalled. "It just tickled us to death." He was particularly proud that the Oregon Trail passed through the north end of the ranch, and he was impressed with the variety and quantity of the grasses, having seen nothing like them during his lifetime in Texas. "We bought some steers, weighed them in and weighed them off the field after several months and their gain was just the same as it would have been in a feed yard. The grass was wonderful." Janet Pollard remembers that her father identified all of the grasses on the property during his first year on the ranch and placed labeled specimens in a glass case.[3]

The trips from Texas to Oregon also provided Clayton with wonderful experiences that he couldn't have in Texas, many associated with food. He delighted in sampling Utah strawberries and Idaho muskmelons, and was particularly impressed with Idaho potatoes. "We stopped to eat in Boise one time and they served you the potatoes that were too large to ship out. They were fantastic. When we first drove out there, I thought there were piles of wood in the fields [but] they were piles of potatoes." Mother-in-law Mernie Graham made pies and cakes on a wood stove during the Oregon blueberry season.[4]

Both J. C. and Clayton's families spent many wonderful summers at the Oregon ranch, but Janet remembers that her father quickly grew bored. "There was simply nothing for him to do. I think this was when he really seriously started writing and doing research about the history of the West. He really loved it." Claytie's impression of his father during this period mirrored his sister's. "He really didn't care about ranching. He was into his history."[5] Clayton said that he first became interested in West Texas history in 1936 when he was asked to write a history of Pecos County as part of the Texas Centennial celebration. "I decided if I had to do a history of Pecos County, I'd first better find out all I could about it," Clayton told a newspaper reporter. Many of his historical resources came from his father's large library, and Clayton also began contacting historians and other scholars for more references.

Throughout the mid-to-late 1950s, Clayton collected information about the history of West Texas in general and Pecos County in particular that he put to paper and published in several venues. Like his father before him, he wrote articles about Pecos County history for the *Fort Stockton Pioneer.* He made substantial contributions to the official program celebrating the Fort Stockton Centennial in 1959.[6] And about this time he began to do serious research for a book-length work that he would eventually expand into three volumes and publish in 1966 as *Never Again,* tracing the history of Texas from prehistoric times to 1961.

As his published work accumulated, Clayton became a sought-after speaker for local schools, clubs, and civic organizations. He was a charter member and served as president of the Fort Stockton Historical Society, and in 1958 he was given a lifetime membership. He was active in several regional and statewide historical organizations in addition to the one in Fort Stockton, including the Pecos County Historical Commission, the West Texas

Historical Association, the Permian Historical Society, and the Texas State Historical Association. He and Chic were a familiar sight at many historical convention meetings throughout Texas. "He had history all over the desks in his office," Janet recalls. "He had history in the basement of his house. Then he had it out in the guest house. He had history everywhere."[7]

Preserving O. W.'s Stories

Another impetus for Clayton to fully engage his love of history was the decision by historian S. D. Myres and the Texas Western College Press to collect and edit his father's stories. The idea to publish the stories in book form had apparently been on Clayton's mind for several years. In the early 1950s he had corresponded with his sister Kathryn, an associate professor of English at Sul Ross State Normal College (now Sul Ross State University) in Alpine, about her desire to publish such a book.[8]

Clayton, despite being slowed now by a light stroke that caused weakness in one of his legs, worked diligently for two years to provide Myres with numerous family letters, photographs, and facts about his father's life and experiences, including a painful recapitulation of O. W.'s dying days. As Clayton ploughed through his father's papers and attempted to make sense of obscure references to events that had occurred seventy or more years in the past, he became aware of how much history is lost after a person dies. "Unfortunately, I have become interested in [O. W.'s] life and his interests at a time he is not available to answer the many questions which now come to my mind," Clayton wrote a friend in 1964. "I think this state of affairs is almost universal among our people, resulting in much of our background being lost to posterity."[9] No doubt learning from this experience, Clayton took pains to record the events of his own life through interviews with family members and others.

Pioneer Surveyor, Frontier Lawyer: The Personal Narrative of O. W. Williams, 1877–1902 was published in 1966, and Clayton was credited on the title page with performing the research for the book. Several months before publication, when he first learned that his research would be recognized, Clayton protested to Myres with typical modesty, insisting that they share credit. "You did the research that was entirely out of my scope and that was more essential to the publication of the book," Clayton told Myres. "You have not taken due credit for your very excellent work on this material."[10]

Clayton was also embarrassed that an article about the forthcoming book in the *Fort Stockton Pioneer* incorrectly attributed the authorship of the book's introduction to him rather than to C. L. Sonnichsen, who at the time of the book's publication was Dean of the Texas Western Graduate School and who published widely in southwestern history and folklore. The article also gave little credit to Myres. "I hope that neither Sonnichsen nor Myres sees the Fort Stockton Pioneer's article," Clayton wrote Carl Hertzog, director of the Texas Western Press. "At no time have I claimed to have written any part of the book, stating only that I have done considerable work on it."[11]

The official "coming out" of the book was celebrated in Fort Stockton on May 20, 1966, with a variety of events, including a graveside eulogy of O. W. at East Hill Cemetery delivered by Sonnichsen; a barbecue attended by 650 invited guests, including many members of the Williams family; and an evening literary program at the First Methodist Church featuring talks by Myres, Hertzog, and Joseph M. Ray, the president of Texas Western College. In a sentimental gesture, Clayton arranged for a quartet to sing two songs at the graveside ceremony, "My Old Kentucky Home" and "Now the Day Is Over." As he explained to Myres, Clayton's father was born in Kentucky and in his old age sang "My Old Kentucky Home" with a "weak and broken voice." The choice of "Now the Day is Over" reflected, in Clayton's mind, that now his father's day was over, a day covering the period from steamboats on the rivers to the atomic bomb.[12] After taking time to write letters of appreciation to the participants after the event, Clayton and Chic traveled to the ranch at La Grande for a much-needed two-month vacation and rest.

Never Again

Three years after the publication of *Pioneer Surveyor, Frontier Lawyer,* Naylor Publishers of San Antonio brought out Clayton's three-volume history of Texas titled *Never Again.* Naylor was a large regional book-publishing house, and company founder Joe Oliver Naylor was known for his support of writers who explored Texas folklore and history. Clayton explained that the book's title reflected that the events that he described "will never happen again" and that he tried "to pick up things that other historians didn't."[13]

Never Again was an ambitious effort by Clayton, but one reviewer, Odie B. Faulk of Oklahoma State University, felt that the effort was perhaps too ambitious. He wrote in *Southwestern Historical Quarterly* that the book revealed little that was new about overall Texas history and in some cases misstated accepted facts. However, Faulk found that *Never Again* did add new information when Clayton concentrated on the history of the Fort Stockton area. "Oral history is blended with family documents unavailable to other scholars to produce some fine writing about the Fort Stockton country," he wrote, noting that Clayton's "chatty informal style" was very readable.[14]

"I'm not a professional writer," Clayton would later tell a newspaper reporter. "I've never had any training in writing. I've always had to have someone go over it and streamline it."[15] Janet Pollard recalls visiting her parents during the summers while she was living in Wisconsin and helping her father with his writing. "I started reading his work. And I would say, 'I don't understand this. Write it again.' He would write his history long-hand and I would type it. Summers I would come home sometimes for three weeks or a month and I would type." Publication of *Never Again* was celebrated in Fort Stockton by a fall book signing in the town's historic Annie Riggs Museum, a presentation of boxed sets of the books to the Pecos County judge and mayor for use in the town's library, and a barbecue hosted by Clayton's son and daughter.[16]

A year after the publication of *Never Again* in 1969, Clayton sold his half-interest in the La Grande ranch to J. C., and the brothers dissolved their legal partnership, though they would remain personally close for the rest of Clayton's life. "J. C. looked up to his big brother fondly," Clay Pollard recalls. "J. C. was not a big part of [the grandchildren's] lives but he was a big part of Paw Paw's life."[17] Janet says that while her father loved Oregon and enjoyed learning about the state's history, he was concerned about being an absentee landowner and also felt some people in the area did not appreciate private property rights. Clayton told his family that he sold because people were breaking the locks on the ranch gates to get onto the land to hunt elk. "I was hot-tempered and thought I might get into trouble by shooting someone."[18] Janet notes that many factors probably influenced the decision to sell, including Clayton's general lack of interest in ranching. But she remembers that her father was sometimes quick on the trigger with firearms, particularly if he felt his family was threatened.

"He wouldn't have shot anyone in Oregon, but when I was growing up in Fort Stockton, a man kept his animals across the little street from our house," she recalls. "One night I heard the door open, then I heard a gunshot and learned that Daddy had shot our neighbor's cow. He thought the cow was an intruder, took it up by the tail, and dragged it back across the street before it died. Another time my cousin, Mary Helen, had come home from college and was staying with us. She had a date and stayed out too late. She said that when she came home, 'Uncle Clayt shot at me. He thought I was an intruder.'"[19]

Throughout the 1970s Clayton continued to focus on his historical research and writing, though he kept his hand in the oil and gas business. His son Claytie was by now heavily involved in what would become a boom-bust-boom career as an oil wildcatter and entrepreneur, and Clayton joined him on some projects. Proud of his son's success, he noted to a friend that Claytie was doing well in oil despite having a degree in animal husbandry. "The principal benefit in any college education is not particularly what course they take, but in how they learn to study and apply themselves to the problem at hand," Clayton said, perhaps reflecting on his own life. "We take advantage of what happens to be available at the time and apply ourselves at that."[20]

Claytie recalls that his father would sometimes ask him to read what he had written during this period. "I was busy growing my business and he would ask me to read something. I felt bad about it but I would say, 'Dad, I'm looking forward, not backward.' And I know that hurt him some. But I think I made him a good son in every other way. He dearly loved his history. And . . . he spent more time I'm sure deleting words to get it small enough than he did putting it all together."[21]

Animal Tales of the West

While he was vitally engaged in writing about the history of West Texas and particularly Pecos County, Clayton did not devote much of his energy to writing about the formation of the area's oil industry. However, he was always willing to share his personal knowledge about the oilfields with others who were writing or researching West Texas oil history, including Myres, his former boss Frank Pickrell, and Berte Haigh, who wrote a history of the University of Texas lands, *Land, Oil and Education,* published in 1986.[22]

Clayton wrote his second major work in 1974 when he was 79 years old. Titled *Animal Tales of the West* and published by Naylor, it was a compilation of stories about animals that expanded some of O. W.'s stories and added both Clayton's experiences with animals and those of government trappers. He enlisted his teenage grandson Clay Pollard to go with him to tape-record the trappers and transcribe the tapes in long hand. "I was a pretty good artist back then, and my grandfather offered to let me draw the illustrations of the animals mentioned in the stories," Clay recalls. "But I was a typical teenager and dragged my feet, so someone at the publishing company did the drawings that were published. I framed and still have the ones that I did." Two drawings of animals by Waldo did appear in the book.

A story Clay told about being attacked by a wounded javelina while hunting ended up in the book, and Clayton let his grandson cosign copies of *Animal Tales* at a book signing in Fort Stockton. A javelina is a peccary, a mammal that bears a superficial resemblance to a pig, and is aggressive enough to injure and sometimes kill humans. "I really didn't deserve to get to do it, but that's the kind of man he was—he wanted to share credit," says Clay, who was pictured in the *Fort Stockton Pioneer* proudly smiling at his grandfather during the book signing. [23] The book was well received.

W. Frank Blair of the University of Texas at Austin, a noted zoologist, described *Animal Tales* as "fascinating for anyone interested in animals or in the largely losing struggle of our large native mammals for survival as the frontier was brought more and more under man's dominion."

Clayton took a hard position in the book on predators, or "varmints" as he termed them, that preyed on cattle and sheep, and he called on Congress to remove restrictions on "poison, cyanide guns, and varmint drives using airplane or helicopters" to control such predators. However, he also emphasized at the end of the book the need to strike a balance between controlling predators that killed off livestock and sheep and "thoughtlessly upsetting nature's balance." Blair praised Clayton for "recognizing the role of predators in control of other species that might become pests."[24]

One story Clayton liked to tell about animals didn't make it into the book. According to Clay, his grandfather said he came upon two rattlesnakes and each one was swallowing the other by the tail. "They were a big circle of snakes. As they consumed each other the circle got smaller and smaller. Paw Paw eventually got bored and killed both of the snakes. He always regretted it. He said he wished he would have waited just to see what happened as those two snakes were swallowing each other."[25]

Horsehead Crossing

During the 1970s Clayton also wrote three articles for the *West Texas Historical Association Yearbook* and two for the *Permian Historical Annual*. The Permian Historical Society named him a fellow in 1976.[26] One of Clayton's favorite historical subjects dealt with the exact location of the famed Horsehead Crossing on the Pecos River, located about 34 miles northeast of Fort Stockton where the river forms the Pecos-Crane County line. Unlike today, the Pecos River in the nineteenth century had considerable streamflow, and Horsehead Crossing was one of the few fordable points on the river.

The crossing first served the Comanche and Kiowa Indians on their raids into Mexico. As they were returning to their base in the plains of eastern New Mexico and northwest Texas, the first water in sixty miles was at the crossing. Legend has it that the name came from the horse skulls that lined the banks on both sides of the river, the remains of dead horses that had drunk too much water. The crossing was used by miners and others who were spurred to head west after the California gold strike in 1848, and later immigrants used the crossing as a place to water their teams before either heading farther west or turning south to Fort Stockton. Subsequently, it became an important stop on the stage route from St. Louis to San Francisco, and beginning in the 1860s cattlemen drove their herds through the crossing. Charles Goodnight and Oliver Loving used the crossing as part of the Goodnight-Loving cattle trail to New Mexico and Colorado. With the coming of the railroads to Texas, the crossing was abandoned and its exact location was forgotten.

In 1936 the Texas Centennial Commission authorized a marker to be placed at what was then thought to be the location of the crossing, about 12 miles northwest of the town of Girvin. The marker was sent to Clayton for placement, perhaps because of his position as a county commissioner, and he "passed the buck" to his brother Waldo to install the marker. Years later, while doing research for an article on the crossing, Clayton attempted to locate its exact position using metal detectors, in the hope of finding artifacts left at campfires, but that effort proved inconclusive. However, in a 1977 article for the *Permian Historical Annual*, which had first been published in 1974 in *Old West* magazine, Clayton used an aerial photograph to show that there were cattle trails leading from the 1936 location of the crossing southwest to Fort Stockton and east-northeast toward an outlet canyon in the Castle Gap Mountains that was part of the Goodnight-Loving trail. The Texas Historical Commission dedicated a new plaque commemorating the crossing in 1974 on Texas FM 11 about two miles to the southwest of the 1936 crossing marker. Clayton and another local historian, Karl Butz, spoke at the plaque's dedication ceremony.[27]

Love for Family

While love for history was a consuming passion for Clayton during the last three decades of his life, love for family was an anchor and a foundation that marked his entire life. At the beginning of *Texas' Last Frontier*, Clayton declared that his family "is the greatest thing in my life" and dedicated the book to Chic, Claytie, and Janet and their families. Claytie believed love for family caused his father to leave the oilfields at the end of the 1920s. "He made a decision—he didn't quite tell me that—but he decided to be a family man, raise a family, and be close to his father, and that's how he ended up staying in Fort Stockton and got into politics."[28] Claytie remembered his father as "always being there" when he was growing up. "In my high school times he was always there. I didn't do for my boys what he did for me. When I played football, he was always there at ballgames. To me, he was always there, at practice, at ballgames. That made a huge difference in my life."[29]

Clayton's supportive approach to raising his own children was in large measure a reaction to his father's stern demeanor and criticism of his children, particularly about financial decisions made by Clayton and Waldo.

HORSE HEAD CROSSING
ON THE
PECOS RIVER

HERE CROSSED THE UNDATED COMANCHE
TRAIL FROM LLANO ESTACADO TO MEXICO.
IN 1850 JOHN R. BARTLETT WHILE
SURVEYING THE MEXICAN BOUNDARY
FOUND THE CROSSING MARKED BY
SKULLS OF HORSES HENCE THE NAME
HORSE HEAD. THE SOUTHERN
OVERLAND MAIL (BUTTERFIELD) ROUTE,
ST. LOUIS TO SAN FRANCISCO, 1858-1861,
AND THE ROAD WEST FROM FORT CONCHO
CROSSED HERE. THE GOODNIGHT-LOVING
TRAIL ESTABLISHED IN 1866 AND TROD
BY TENS OF THOUSANDS OF TEXAS
LONGHORNS CAME HERE AND TURNED
UP THE EAST BANK OF THE PECOS
FOR FORT SUMNER AND INTO COLORADO

Erected by the State of Texas
1936

Clayton with the original 1936 state historical marker commemorating Horse Head Crossing on the Pecos River, which served Indians and later white settlers as a fording point. Clayton devoted much energy later in his life to pinpointing the exact location of the crossing, and his efforts resulted in a new plaque that was dedicated in 1974 southwest of the 1936 marker.

Writing from the solitude of the Oregon ranch in 1960 to his niece Mari Helen Schultz, Clayton reflected on the relationships between parents and their children:

> I only wish that my father, although proud and loyal to his children but evidently disappointed in their achievements, could now see his grandchildren and great grandchildren. Waldo had the brains but not the will for worldly achievements. God rest his soul! He held a premonition that fate was against him and

even at that it would have been a happier life for him if he had never had such an idea. It is better never to know the misfortunes ahead of you! What a noble figure and grand spirit he could have shown had fate been such that Freedom could have been his to exploit his hobbies?

And so it is today with our children. What is their future? Can they pick out some occupation for their future which they will love and enjoy and at the same time make a remunerative living? We cannot pick it out for them, and they will be out in the cold world before they know what they want to do; and the necessity of doing may never allow them the privilege of doing that which is their love and most acceptable occupation.[30]

Janet says she always sensed her father's love and respect for his mother and father. "He said, 'I never pass their house that I don't get a lump in my throat,'" she recalls. But she remembers that her father, unlike O. W., was supportive rather than critical. "What he took from Judge was to tell us to pay attention to your surroundings," Janet says. "He didn't think in a box and he didn't have us think in a box."[31] Claytie believes that his father's finest trait was his integrity, particularly when it came to dealing with other members of the Williams's family. "I think the thing that I am most proud of about my dad, among many things, was his integrity. When my grandfather died, he stepped in and represented the estate of O. W. Williams. There were all these heirs and he had many chances to screw them and he never did. He was totally honest."[32]

As Claytie and Janet married and began families of their own, Clayton and Chic became a loving presence in the lives of their nine grandchildren. Frequently, a grandchild would get to spend a summer in Fort Stockton and share the life of a West Texas rancher, learning to ride first on burros and then horses, clearing brush, hunting rabbits and later deer, and camping under the stars. Clayton passed along with these experiences many lessons about life. Clay Pollard remembers how the family gathered at holidays and how his grandfather loved to tell stories. "Paw Paw would sit at the head of the table and tell stories the whole time. He and Chic would be the center of attention. Then as a family we would watch whatever football game was going on, usually Texas A&M if it was Thanksgiving and the Dallas Cowboys if it was Christmas. These events cemented the family bond and gave us all even more appreciation for Paw Paw and Chic."[33]

After the death of O. W. in 1946, Clayton assumed the role of a loving family patriarch. He managed O. W.'s extensive estate and dispensed annual payments to the heirs, refusing for several years to take the full estate management commission to which he was entitled. His grandchildren, nieces, and nephews routinely asked for his advice about business and financial matters, and he always took the time to respond either in writing or with a timely telephone call, offering his assistance wherever possible.

Fort Stockton became "home" for many members of the Williams clan, no doubt in large measure to be closer to Clayton and Chic. His sister Kathryn, who died in 1986 after retiring to Keota, Oklahoma, visited frequently while living in Alpine. Ermine, who had acquired her siblings' inherited interests in O. W.'s house, began extensive renovations before a stroke cut her life short in 1950. One of Ermine's three children, Sarah Garnett, ultimately inherited the house and lived there from 1965 until 2003, serving as an Episcopal priest for twenty-three years. The Williams family ultimately owned the house for more than 100 years.

Chic's brother Oscar Graham and his wife Mary, known to the family as Aunt Mary E., built the house across the street from Clayton and Chic. "Uncle Oscar and Aunt Mary E. were very close to Mother and Daddy," Janet says. Mernie, Chic's mother, lived with her daughter and son-in-law from shortly after Claytie was born in 1931 until her death in 1972. J. C., who had lived and worked in China, Puerto Rico, India, Australia, and New York, retired with his wife to Fort Stockton in 1964, and he and Clayton swapped stories and reminisced over coffee several mornings a week.

Family Stories

One trait that Clayton never lost, even into his eighties, was his quick temper and willingness to fight. "Daddy was a quiet, dignified, and gentle man, but if someone crossed a line, he could get angry pretty quick," Janet says. "And, even at an advanced age, he could get physical." Clay Pollard recalls several stories about his grandfather's quick temper, including one incident that occurred when his older brother, Scott, was married in Dallas in 1979.

"The wedding party was staying at the Stoneleigh Hotel and I was best man. So I was real busy, doing all the things that a best man had to do. On the afternoon of the wedding, I was passing through the lobby when I saw Paw Paw leaning nonchalantly over the counter of the reception desk like he was having a conversation with the clerk." The clerk, Clay recalls, had apparently been rude to several members of the wedding party during their time at the hotel and his grandfather had noticed the behavior. "Paw Paw gets my attention and says, 'Clay, come here.' I walked over to him and he said, 'See this man. He's on the telephone calling the police. He says that I tried to hit him. I didn't try to hit him. I reached over and tried to take his glasses off so I could hit him.' I realized that the police were probably going to come for my eighty-four-year-old grandfather so I grabbed him by the arm and said, 'Come on, Paw Paw, we got to go.'"

Scott Pollard adds, "When they got Paw Paw out of the room, they gathered up my groomsmen. I had three or four groomsmen who were six-foot-two or better and they put him in the center of them so that they could get him out to the parking lot without being seen. They had him well hidden." Clay and Janet both recall that Clayton danced every dance at the wedding reception later that evening. The next morning, according to Janet, her father went down for breakfast at about six o'clock and met her cousin Harriet Herns. "He and Harriet were visiting and he told her he had come to Dallas a few days before the wedding. She said, 'What for, Uncle Clayt?' He said, 'Oh, to get a physical.' And she said, 'Any man who can threaten a desk clerk, dance all night, and get up at six in the morning I declare healthy.'"[34]

Clay Pollard recalls that four years earlier his grandfather had almost gotten into another altercation at a high school football game. "Scott was a senior in high school and played on the varsity football team. Grandma Chic and Paw Paw came over to watch Scott play. It was a cold night and as Paw Paw was getting older, we were more conscious of trying to keep him warm so he wouldn't get sick." Clay says that his grandmother had knitted a red wool cap with a ball on the top that she insisted Clayton wear to the football game. "We walked toward seats on the fifty-yard line and people stood up to let us by. My dad and mom were first in line, Grandma Chic was behind them, and Paw Paw was right in front of me.

"So as we passed one gentleman, he looked up and saw Paw Paw in that red cap—he did look kinda funny with an overcoat and a red wool cap—and this guy kind of snickered when he saw Paw Paw. Paw Paw looked at him and said, 'You son of a bitch! Do you want me to kick your ass?' I saw the fight was just about to break out, and I pushed Paw Paw on the back and moved him on down to where we were going to sit. The guy kept his mouth shut the rest of the game."[35]

About twenty years after his grandfather died, Clay Pollard was at Claytie's deer hunting camp. Also at the camp was former Pecos County Commissioner Bill Moody, who Clay believes was about eighty years old at the time. "We were sitting there enjoying campfire conversation and I asked Bill to tell me a story about my grandfather. Bill said that one day he was outside the Pecos County courthouse when the court was taking its break. Two opposing attorneys were out in the front yard fussing at each other. Bill said my grandfather went over to play peacemaker and tried to stop the two men from arguing so it wouldn't escalate. Well, they turned on him. Bill said he whupped them both."[36]

Clay says his grandfather believed that daily exercise would enable him to live a long and healthy life, like his father, who lived to ninety-four, his grandfather, who lived to ninety-eight, and his great grandfather, who lived to ninety. Clayton tried to walk a mile a day even into his late eighties. "He would be out the door for a walk at 5:30 in the morning to avoid the heat," Clay recalls. "When he began to worry about stray dogs or the occasional transient who would hit him up for money, he carried a cane for self-defense. I wouldn't have messed with him."[37]

Clayton and Chic

Janet says that her mother and father had a wonderful and loving relationship, and when they did have an altercation, it was more "cute than angry. Seemed like a lot of Christmases my mother couldn't wait for Christmas to be over and she would say, 'Now we are going to save money.' Well, we hardly spent any, but she really liked to hold onto that money. So one year, when I was married and living in Midland, she showed up at my door and said she was going to spend the night. And I thought, 'How nice. My mother is going to spend the night.' A few hours later my father was on the telephone, asking to speak to Mother. They had had a fight. The lady who ran the library in Fort Stockton was fundraising for the library and he had given her some money. Mother thought he had given too much and I also think she was just a little bit jealous. I was thirty-two years old and had never once heard my parents fuss. I just put my head down and cried. But the argument ended quickly and mother went home the next day."[38]

Clay Pollard recalls another example of their grandparents' "fun" relationship during a car ride in the early 1970s. "My Uncle Claytie was in the middle of his drilling operations and pipeline construction in the Giddings field. He had a big barbecue in Giddings and I'm driving Scott, Grandma and Paw Paw to Giddings for this barbecue. On the way over there I'm looking in my rearview mirror at Paw Paw and Chic in the backseat. One moment Chic would reach over and kinda push Paw Paw's hair down. Then the next moment she'd reach over and kinda straighten his collar. And the next moment she'd try something else. So Paw Paw started making all these expressions with his face, moving his hands, and saying he couldn't do anything right and that Chic was always fixing this and fixing that. So Scott and I got a good laugh out of that."

Clayton and Chic enjoyed driving over to Midland to eat at a Chinese restaurant named the Blue Star Inn, and they would frequently treat one or more grandsons to dinner. On one occasion, Chic was driving and pulled into a parking space at the restaurant with the car straddling the lines that defined the space. "Grandpa Clayt told Chic to back up and pull the car in straight. She backed up, pulled in again, and was even more over the line. All he could say was, 'Damn, Chic.'"[39]

Janet notes that her parents' first date had been on a dance floor and that the couple continued to enjoy dancing all of their lives together. "They belonged to the 66 Dance Club and went to dances once a month. I guess it was named 66 for the number of people who belonged to it. He would get one or two swigs of apricot brandy before he left the 66 Club. He also loved to play dominoes every night with his mother-in-law, Mernie. She would throw the dominoes at him and he would just die laughing. He had a good spirit. He was a happy man."[40]

In the fall of 1978, Clayton and Chic celebrated their fiftieth wedding anniversary at a party given by Claytie and Janet in Fort Stockton. "Claytie paid for it all and I did the work," says Janet. "We hired a hall, ordered a huge fancy cake, and invited as many family and friends as we could think of. And would you believe that it rained that night and still a thousand people showed up? It was a beautiful celebration and they were so sweet together, dancing to the music of a wonderful band from Dallas."[41]

The Last Chapter

As Clayton entered his mid-eighties, he increasingly began to have bouts of bad health. The West Texas dust storms had taken a toll on his lungs, and a severe case of pneumonia sent him to Claytie's Happy Cove Ranch near Alpine to recover and escape the dust. He returned to Fort Stockton and spent the summer of 1981 engaged in rewriting and polishing the manuscript for his final book, *Texas' Last Frontier: Fort Stockton and the Trans-Pecos, 1861–1895,* which was published in April of 1982 by Texas A&M University Press. The book was edited by Ernest Wallace, Horn Professor of History, emeritus, at Texas Tech University.

Grandson Scott Pollard thinks that the use of the word "frontier" in the title was particularly appropriate. "The land was harsh and few saw reason to settle here until nothing else was left," says Scott. "It seldom rains and there is little water here. A good rain will change the mood of the entire community. Everything that grows is covered in thorns. If an area isn't covered in rocks, the slightest step will kick up dust. The people who settled this land had to be tough. They also had to be resourceful to scrape out a living. They were open to new technology because it helped them tame this wild land."[42]

In the preface to *Texas' Last Frontier,* Clayton laid out his reasons for writing the book:

Chic and Clayton dance at their fiftieth wedding anniversary in Fort Stockton in 1978. More than a thousand people attended the celebration.

A jolly Williams family gathers in 1982 to celebrate the publication of Clayton's Texas' Last Frontier: Fort Stockton and the Trans-Pecos 1861–1895. *Front left are Clayton's grandsons Jeff and Clayton Williams and Adam Pollard. Behind them are grandson Clay Pollard; Clayton and Chic; granddaughter Kelvie Cleverdon; daughter Janet Williams Pollard; granddaughter Allyson Groner; and daughter-in-law Modesta, her husband Claytie, and their daughter Chim Welborn.*

Fort Stockton and the surrounding area, I believe, epitomize the American frontier, and local people, still noticeably characterized by their frontier heritage, long to have that heritage recorded for their posterity. It would be a rare person indeed who could examine the historical markers in and around Fort Stockton, the old guardhouse, the barracks, the Oldest House in Town, and the irrigation ditches without being overwhelmed with a feeling of admiration for the stamina and optimism of the pioneers they commemorate.[43]

Reviewing the book for *Southwestern Historical Quarterly,* Brad Agnew of Northeastern State University said it lacked a "coherent conceptual framework" and would have benefited from tighter editing. However, he found Clayton's work "particularly rich in anecdotes and stories passed down from the pioneers who settled the trans-Pecos frontier," and said that *Texas' Last Frontier* "makes a significant contribution to the literature of the Lone Star State."[44]

Writing in the *Texas Aggie*, Herbert H. Lang, A&M professor of history, said the book "is a fortunate blending of careful research and sound scholarship. The contents are well-organized. The style is spritely, and descriptive passages are outstanding. . . . The writing of local history has always been the responsibility of the non-professional historian. In this world of change it is reassuring to know that the great tradition is being maintained by men like Clayton Williams, and that the Texas A&M University Press recognizes that skillful and dedicated historians produce first rate history of lasting significance."[45] Judith Parsons of Sul Ross State University wrote that Clayton and Wallace "have produced an exceptionally fine work of local history. Local history is all too often inadequately researched, narrow in scope, and shabbily produced. Williams avoids all of these pitfalls."[46]

A few months before *Texas' Last Frontier* was released, Clayton and Chic spent a month in Honolulu. Fearing that another bout of pneumonia might strike his eighty-six-year-old father, Claytie paid for the trip. Clay Pollard recalls that his grandparents were so homesick for their family that they

Clayton signs copies of Texas' Last Frontier *at the Haley Memorial Library and History Center in Midland as Texas author and historian J. Evetts Haley and Chic look on.*

paid for the grandchildren to come to Hawaii. "Not everyone could come, but many of us did. There were so many that whenever we went anywhere as a group, the crosswalk light would change before we could all cross the street."[47]

That fall Clayton and Chic celebrated their fifty-fourth wedding anniversary with a trip to New York and then to Halifax, Nova Scotia, the city from which Clayton had sailed to war in 1917. "It was a happy trip," Janet recalls, "full of reminiscences." After they returned, the West Texas Historical Association dedicated its 1982 *Yearbook* to Clayton, and the family gathered to celebrate at Thanksgiving. They spent Christmas in Midland at the home of Claytie and his wife Modesta, where Clayton handed out his yearly Christmas checks to the grandchildren. After the New Year, daughter Janet began noticing that her father seemed to be more tired than usual, though he kept up a busy schedule with frequent trips to historical meetings and family events. He also completed substantial contributions to a two-volume *Pecos County History* that would be published by the Pecos County Historical Commission in 1984.

In the spring of 1983 Clayton and Chic attended Kathryn's ninety-first birthday party at her son's ranch on Kerr Lake near Keota, Oklahoma. The event featured horse races at the nearby town of Sallisaw and a fish fry, a hayride, and a dance with live country music. "My father danced every dance," Janet says. "He loved music and he loved dancing." Kathryn's daughter Harriet recalls seeing Clayton up early one morning in the lobby of the modest hotel where he and Chic were staying during the party. "He was seated in a chair in the lobby with a large group of younger relatives seated around him on the bare floor as he told family stories. He was alert and in good health then." Two months later Clayton was hospitalized with a severe case of hiccups, which the family later learned was a symptom of cancer around the diaphragm. "Daddy was always cheerful, but he was also always tired," Janet recalls. "He tried to work on his autobiography at the start of the summer but just didn't have the energy. And he was losing a lot of weight. Then he had to have a growth removed and it was cancer. After more tests, we learned in July that he had terminal lung cancer."[48]

Clayton took his diagnosis with his usual quiet dignity and strength. "He simply said, 'It's OK. All is well with my soul,'" Janet recalls. As word of his illness spread in the Fort Stockton community, people began calling and coming by his home, bringing flowers and food. J. C. brought his brother roses every day. Janet moved into Clayton's home in Fort Stockton while her husband Bob commuted to and from Midland. Claytie and his wife Modesta moved to their home in Fort Stockton and Claytie ran his businesses from there.

Claytie paid for a local restaurant to bring in enchiladas every day, and he had ice cream, his father's favorite treat, brought in from Midland. He also paid for the grandchildren who lived far away to fly in to visit their grandfather. "These were expensive gestures, but so like Claytie," says Janet. By now Clayton was on oxygen and using a wheelchair, and he began putting his life in its final order. "He checked on his will, checked on mother, checked on everyone else," Janet recalls. "Claytie and I started bringing our children home to visit with him. It was kind of like what you read in the Bible—he was the elder giving his blessing to everyone in the family."[49]

Despite his worsening health, Clayton accepted an invitation from his son to attend Claytie's annual cattle sale and party in August at his ranch in Alpine. Over a period of several years, Claytie had developed a successful purebred cattle business at his farm outside Fort Stockton, specializing in Brangus, a cross of Brahman and Angus. The annual sale, followed by a spectacular party attended by several thousand people, was held at Happy Cove Ranch.[50] "Daddy said, 'Of course I will go,' Janet recalls. 'I might as well be there at that pretty ranch.'"

Claytie flew to Fort Stockton in his company helicopter to fly his father, mother, and Janet to Alpine about a week before the sale and party. "Daddy looked down on the country as we passed over it and he would tell stories, about this water well being here and that ranch being there," Janet says. "We flew down low over the farm that he created by drilling those deep water wells, and then over the dry country, seeing that one side of the road was devoid of grass and the other side of the road had grass. We just talked about rain and ranching and oil."

As the helicopter approached Alpine, Clayton asked the pilot to fly low over Black Canyon. "Daddy said, 'Fly down close to those springs,'" Janet recalls. "He said, 'My dad and I used to camp out there on a buckboard on our way to the Big Bend when he surveyed in the summers.' That was such an interesting piece of history to me. Here we were in a helicopter looking at where he used to be in a buckboard."

When the helicopter arrived at the ranch about noon, a group of ten men on horseback were waiting. Included among them were Janet's son Adam and Claytie's sons Clayton Wade and Jeff. Claytie had the men hide in a draw from which they communicated by radio with the helicopter pilot. "As we landed, the cowboys came out and started riding around the helicopter in a circle," Janet says. "When we opened the door, the men took their hats off and put them over their hearts. It was a heartwarming and beautiful tribute, and Daddy just loved it."

Once Clayton and Chic settled in at the ranch, they developed a routine that included swimming every day in the large pool that Claytie had built in the shape of a boot. "Daddy loved water from growing up around Comanche Springs, so we would all go out and swim with him. He would just float and talk and it became kind of a priority of the day. We would look up at these beautiful mountains and we would quote Daddy's favorite verse from the scripture—"I will lift up mine eyes unto the hills, from whence cometh my help." A few days before the ranch party, it rained. "It rained so hard that the portable potties brought in for the party started floating down the creek," Janet recalls. "But the rain was such a blessed relief that everyone was joyous. Suddenly Alpine looked like Scotland."

As the date for the sale and party approached, family and friends whom Clayton had known all his life began arriving. One guest who wanted to meet Clayton was author James Michener, who had become friends with Claytie while doing research for Michener's historical novel *Texas* that was published two years later. "I remember being in the room when Michener came in," Janet says. "He said to Daddy, 'I want to thank you, Mr. Williams, for all of the books you have written and researched and the history that you have left me to work with to write my book.' Michener visited for about an hour. Daddy didn't read fiction but he was very pleased and knew who Michener was."[51] With Michener's encouragement, after his father's death Claytie donated $150,000 to Texas A&M University Press to help the Clayton Wheat Williams Texas Life Series support the publication of books about daily life in Texas during different periods of the state's history.[52]

On the day the auction was to take place, Clayton was carried down and introduced to the crowd. Both Janet and Claytie were "choked up," so Modesta made the introduction. When Clayton tipped the Stetson hat he was wearing, the crowd gave him a standing ovation. After the party ended, Claytie and Modesta returned to Midland and Janet's husband Bob left the ranch for a business trip. Clayton's condition worsened and he suffered from considerable pain. Janet gave him opium orally to control the pain, but one night her father stopped breathing. "I didn't know the first thing about CPR. I grabbed his back, set him up and started beating his chest, and he started breathing again.

"While I was going through all this, a man whom Claytie had brought back from one of his big-game-hunting trips to work at the ranch was standing nearby. He had grown quite fond of Daddy and was very attentive, helping him out with lots of daily tasks. When he saw me beating on Daddy's chest, he started yelling, 'You've killed Papa! You've killed Papa!' I told him that if he didn't shut up I was going to kill him. I knew then that we had to hire a nurse."[53]

Clayton agreed to return to Fort Stockton, and the next day Janet called Claytie and asked him to send the helicopter to take them home. "Daddy was going down further and further and got to where he wouldn't open his eyes," Janet recalls. "But when we got close to home he opened his eyes and said, 'I see Fort Stockton.' The helicopter landed near his house, and many of the members of the local VFW (Veterans of Foreign Wars) met us and carried Daddy into his home."[54] Clayton lived for another two weeks. During his last days Claytie suggested that the family have a fish fry at Clayton's Fort Stockton home. "It was a great idea," says Janet. "All of the people who loved him could come and say goodbye." However, it was obvious to those attending that Clayton was in considerable pain.

On the day before his death, Clayton complained of the pain. "Daddy said, 'I can't take much more,' Janet recalls. "It was a hard day." Clayton Williams died on the morning of September 9, 1983, one day before his and Chic's fifty-fifth wedding anniversary. Claytie and Modesta had returned to Midland just the day before. "I couldn't stand to watch him die," Claytie says.[55]

Clay Pollard was attending classes at Texas A&M when his grandfather died, but recalls that he had visited him just a week earlier. "The last weekend I went to see him he was laid up in bed. You could tell that he didn't

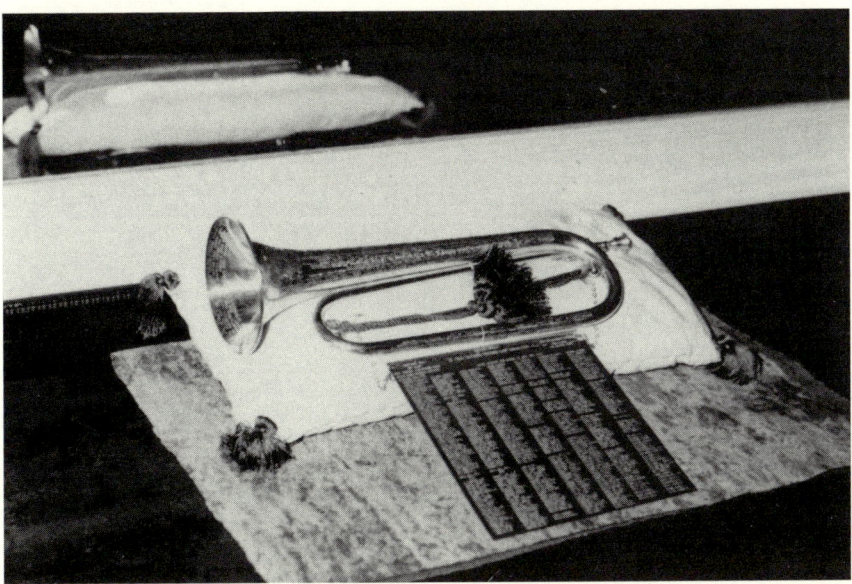

This silver bugle, part of the permanent collection in the Alamo in San Antonio, commemorates the "90-day wonders," men who trained with The First Campers during World War I. After he died in 1983, Clayton's name was inscribed on the bugle along with the names of other campers who had died. His name is also included on a plaque that honors The First Campers on the battleship USS Texas, which is moored at the San Jacinto Battleground State Historic site near Houston.

have long. He was lying there peacefully with his eyes closed. I hugged him just like I was a little boy even though I was a grown man. I told him I was going to marry Jeanne. He knew Jeanne. And even though I hadn't talked with Jeanne about it yet, I knew she was going to be the girl that I would marry. My grandfather's last words to me were 'Bless your heart.'"[56]

Clayton's death was reported extensively by television stations in Midland and Odessa and by local and regional newspapers. The funeral was held at the Fort Stockton United Methodist Church. Clay remembers that the family wanted to keep his grandfather's Texas A&M class ring and not have it buried with him, but they also wanted him to be wearing the ring at the funeral service. "We didn't want to chance losing his personal ring, so I took off my own A&M ring and placed it on his finger while he lay in state. We just couldn't imagine my grandfather without an Aggie ring." From that day on, Chic would remind Clay that his grandfather had worn his A&M ring.[57]

"After my grandfather died, Chic would always be happy and bubbly whenever she was around her kids and grandkids, but whenever she thought of Paw Paw, she would be sad," Clay says. "She always pined for Paw Paw. She never got over losing him."[58]

Janet remembers her father's last hours. "I don't know how you get that premonition that a person is going to die, but there just becomes a kind of stillness over everything as the angels are there preparing him to go. That night we had a nurse, and mother and I were sleeping in the bed next to him and I heard his breathing change. I got mother up and we called Aunt Mary E. from across the street to come over. We stood around and prayed and wished him *vaya con Dios.*"

Epilogue

In the summer of 1984, Clayton Williams was nominated to the Hall of Fame of the Permian Basin Petroleum Museum in Midland. He was formally inducted in 1987, the first year he was eligible after his nomination. There he joined Carl Cromwell, Frank Pickrell, and others who had "made outstanding contributions to the development of the petroleum industry or have served as worthy examples to those in the petroleum industry." Eighteen years later, Claytie would join him in the Hall.

Included in the documentation accompanying Clayton's nomination letter was a long recitation of his personal and professional achievements. The list of accomplishments described a lifetime spent as a successful oilman, rancher, farmer, businessman, civic leader, and historian. But as is often the case in a factual description of a person's accomplishments in life, the inner man was not described—the man that his family, friends, and colleagues remembered so well—his traits of strength, dignity, honesty and, perhaps above all, his love of family.

He was a talented self-taught petroleum geologist, but Williams chose family over a life in the oilfields. Janet Pollard remembers that he was always close by as she was growing up.

> Every day he came home for lunch and we ran to meet him and give him a hug even though we had just seen him that morning. He would come home with funny stories.
>
> I never heard my dad complain about working hard. He looked toward the positive side. He always seemed to look ahead and never back.
>
> When I was a tot I remember that I didn't go to sleep until I held his thumb through the crib, then I would go to sleep. We had a breathtakingly wonderful childhood, Claytie and I.

Daddy was kind of like the 23rd Psalm. He was a shepherd and a mentor and a shelter.

Delivering the eulogy at Williams's funeral service, Dr. O. A. McBrayer told a story about a conversation he had with Clayton in the last days of his life. McBrayer had asked Williams how he would like to be remembered. Taking several seconds to think before replying, Clayton said, "I'd like to be remembered as a kind and caring man."[1]

Claytie and I were very fearless while growing up and part of it was because Daddy wasn't fearful. He was the kind of mentor who did not tear up your self-respect while he directed you. He didn't take away your good feelings about yourself. Daddy was not sarcastic. I loved that he did not downgrade us in any way. Nor did I hear him downgrade many other people. I appreciated being raised by two parents who were not critical. To have parents like that is a blessing.

Lifelong friend Frank Baker once commented on Williams's dying wish to be remembered as a kind and caring man. "It was a rare insight, indeed, to realize that being considered a man who was a brother to so many other members of the human race, that being considered caring and kind, meant more in the long run than any of his many accomplishments—to him, to his family, and to all who knew and loved him.

"He served his community, county, state, and nation well. He cast a larger-than-life shadow on the world around him, while always coming across to any who knew him as a kind and caring man."[2]

Clayton was buried in an oak casket in Fort Stockton's East Hill Cemetery alongside his father and mother. He would be joined there four years later by J. C. Chic died in 1998 at the age of 92. At the top of Clayton's tombstone are the beginning words of the Twenty-third Psalm, "The Lord is My Shepherd," followed by a simple summing up: "A Native Son, Patriot, Officer World War I, Pioneer Petroleum Engineer, Rancher, Historian, Writer." At the bottom of the stone is carved "A Kind and Loving Man."

"Be strong and courageous," my dad said to me. Following him, could I be any less? Daddy, your book is finished.

To Daddy with Love,
Janet

Notes

Chapter 1

1. Robert H. Schmidt, "Trans-Pecos," *Handbook of Texas Online*, http://www.tsha online.org/.

2. "Midland-Odessa Combined Statistical Area," *Wikipedia*, last modified July 26, 2010, http://en.wikipedia.org/.

3. "Fort Stockton, Texas," *City-Data*, accessed August 26, 2010, http://www.city-data .com/city/Fort-Stockton-Texas.html.

4. There is no single source for this popular expression, which has been used to describe Texas (and other states), the cattle business, and the hard life of women living on ranches.

5. O. W. Williams's writings are collected in *Pioneer Surveyor, Frontier Lawyer: The Personal Narrative of O. W. Williams, 1877–1902*, ed. S. D. Myres, with an introduction by C. L. Sonnischen (El Paso: Texas Western College Press, 1966). Clayton Williams assisted in editing this book. It collects O. W.'s writings about his experiences in Texas and New Mexico as a surveyor and silver miner over a twenty-five-year period that encompassed most of the last quarter of the nineteenth century. O. W. published the stories as pamphlets primarily to teach his children and grandchildren about the early history of this part of the Southwest. Unless otherwise noted, our understanding of O. W. Williams's early life and career is taken from this book. However, many of the stories describing the lives of O. W.'s ancestors have been compiled and, for the most part, documented by O. W.'s sister, Jesse Williams Hart of Los Angeles. See Mrs. A. E. (Jesse Williams) Hart, "The Callaway Family of Virginia" (unpublished manuscript, 1929), and "Col. Richard Callaway and Allied Families," 2 vols. (unpublished manuscript, 1934–35), both in the authors' possession.

6. O. W. Williams to J. C. Williams, June 16, 1918, Oscar Waldo Williams Papers, Haley Memorial Library and History Center, Midland, Texas (hereafter cited as Haley Library). Included among the Haley Library papers are twelve volumes of letters written from 1894 to 1936 that were transcribed from the handwritten original letters, duplicated, and given by Clayton Williams to the Center for American History at the University of Texas at Austin, where they were examined by Louis Gwin. Hereafter, letters to and from O. W. Williams will be designated by the initials "O. W. W." and letters to and from J. C. Williams and Clayton W. Williams will be designated "J. C. W." and "C. W. W."

7. Adam Pollard, e-mail message to Louis Gwin, September 7, 2009.

8. Clayton Williams's published books are *Never Again,* 3 vols. (San Antonio: Naylor, 1969); *Animal Tales of the West* (San Antonio: Naylor, 1974); and *Texas' Last Frontier: Fort Stockton and the Trans-Pecos, 1861–1895* (College Station: Texas A&M University Press, 1982).

9. C. L. Sonnischen, intro., *Pioneer Surveyor, Frontier Lawyer,* 3–4.

10. Mike Cochran, *Claytie: The Roller-Coaster Life of a Texas Wildcatter* (College Station: Texas A&M University Press, 2007). Clayton W. Williams Jr., known as Claytie, has periodically appeared on the *Forbes* list of the 400 wealthiest Americans. In addition to oil and natural gas exploration, his business interests have included gas pipeline operations, ranching, real estate development and management, banking, and telecommunications. He is chairman of the board, president, chief executive officer, and a director of Clayton Williams Energy, Inc., with corporate offices in Midland, Texas. He ran unsuccessfully in 1990 as the Republication candidate for governor of Texas.

11. Scott Pollard, e-mail message to Louis Gwin, August 2, 2009.

12. Ibid.

13. Sonnischen, intro., *Pioneer Surveyor, Frontier Lawyer,* 5.

14. According to Hart, the Colyer name was also spelled Colyar and Collier.

15. Hart, "Col. Richard Callaway," 2:553.

16. Ibid. As documentation, Hart reproduces the original letter dated July 29, 1926, from the War Department, stating that its records "showed one John Collier served as a private in Captain Thomas Hill's Company, 5th Virginia, known also as the 5th and 11th Virginia regiment, commanded by Colonel Wm. Russell, and in Captain Thomas Hill's Company, 7th Virginia regiment of foot, commanded by Colonel McClenahan, Revolutionary War" from February 13, 1777, to November 1779. The War Department also said that "a man named John Collier served in the Revolutionary War in Captain James Knox's company, 8th Virginia regiment" from May 25, 1776, to April 30, 1777.

17. Ibid., 560–61; and an unfinished autobiography by Clayton Williams, in the authors' possession. This is a manuscript that Clayton began after *Texas' Last Frontier* was published in 1982. In it he wrote that Colyer hid in a hollow log and that his Indian pursuers sat on the log for a time while Colyer was inside.

18. Wilbur Edel, *Kekionga! The Worst Defeat in the History of the U.S. Army* (Westport, Connecticut: Praeger, 1997), 82; Harvey Lewis Carter, *The Life and Times of Little Turtle: First Sagamore of the Wabash* (Urbana: University of Illinois Press, 1987), 100–111; and Hart, "Col. Richard Callaway," 2:560–61.

19. Hart, "The Callaway Family," 222. Citing records provided by the Bureau of Pensions, Hart said that "Ensign" Jesse Williams served as a private, a sergeant, and an ensign with Maryland troops from 1776–79. There is no evidence that he served in 1780 when he moved his family from Maryland to Virginia. He then reenlisted in 1781 as a private with Virginia troops. He was overcome by heat in 1835 while supervising slaves on his tobacco farm in Rockcastle County, Kentucky, and died at age eighty-five.

20. Hart, "Col. Richard Callaway," 2:535–42; and O. W. W. to J. C. W., September 9, 1923, Haley Library. O.W. said that the year of the journey was "1825+1826."

21. Hart, "Col. Richard Callaway," 2:535.

22. Ibid., 1:92–97. The trail ran for 200 miles from Sycamore Shoals on the Watauga River in far eastern Tennessee, through the Cumberland Gap, to the site of Boonesborough, first called Fort Boone, on the south side of the Kentucky River near its junction with Otter Creek in eastern Kentucky.

23. Meredith Mason Brown, *Frontiersman: Daniel Boone and the Making of America* (Baton Rouge: Louisiana State University Press, 2008), 106–10; and Hart, "Col. Richard Callaway," 1:341. Hart said (although without documentation) that while "there is a tradition" that Holder died in an Indian ambush, he actually died of natural causes at his home in Clark County, Kentucky. Hart describes the life of Catherine Holder, her grandmother, in "The Callaway Family," 221 and "Col. Richard Callaway," 2:438.

24. Hart, "Col. Richard Callaway," 2:535; and O. W. W. to J. C. W., March 8, 1931, Haley Library. In O. W.'s retelling of the story, Richard Gott Williams purchased horses and mules in Richmond, Kentucky, and sent them overland to St. Louis. There he procured twelve Conestoga wagons, loaded them with goods, and took his caravan on the Boone road up the Missouri River, crossed at St. Charles, and moved along the north side to Franklin, Missouri. There he crossed again to camp on the south side.

25. O. W. W. to J. C. W., September 9, 1923, and March 8, 1931; Hart, "Col. Richard Callaway," 2:508; and C. L. Sonnischen, intro., *Pioneer Surveyor, Frontier Lawyer*, 5–6. O. W.'s story of his grandfather's trip on the Santa Fe Trail is retold in Clayton Williams's *Never Again*, 2:17–20.

26. Carson's story about the accident is found in his autobiography. See Milo Milton Quaife, ed., *Kit Carson's Autobiography* (Lincoln: University of Nebraska Press, 1966), 5–6. Many others have retold the story. See, for example, Hampton Sides, *Blood and Thunder: An Epic of the American West* (New York: Doubleday, 2006), 11–12. Janet Pollard recalls hearing the story of Broadus's accident told many times during her youth by members of her family.

27. O. W. W to J. C. W., September 9, 1923.

28. Louise Barry, "Kansas before 1854: A Revised Annals," *Kansas Historical Quarterly* 28, no. 1 (1962): 29.

29. Hart, "Col. Richard Callaway," 2:535–36; Sonnischen, intro., *Pioneer Surveyor, Frontier Lawyer*, 6; O. W. W. to J. C. W., September 30, 1923, Haley Library; and Williams, *Never Again*, vol. 2, 19–20.

30. Hart, "Col. Richard Callaway," 2:535–36; Sonnischen, intro., *Pioneer Surveyor, Frontier Lawyer*, 6; and Hart, "Col. Richard Callaway," 2:439–40.

31. James W. Covington, ed., "A Robbery on the Santa Fe Trail, 1827," *Kansas Historical Quarterly* 21, no. 7 (1955): 560–63.

32. Hart, "Col. Richard Callaway," 1:271.

33. Ibid., 2:487.

34. Ibid., 1:263, 276. According to Hart, Jesse Caleb moved to Illinois primarily because he wanted to live in a state that had outlawed slavery. Hart wrote that a young slave woman was given to Mary Collier upon her marriage and lived with the family for seven years. When the family moved, Jesse Caleb freed the slave, even though she asked to remain with the family. Many years later Mary Collier, during a visit to Kentucky, encountered the woman, who was purported to have said, "Oh Miss Mary, Miss Mary, why didn't you take me with you?"

35. Ibid., 1:267.

36. Sonnischen, intro., *Pioneer Surveyor, Frontier Lawyer*, 6; Hart, "Col. Richard Callaway," 1:263; Mark Sorensen, president, Illinois State Historical Society, e-mail message to Louis Gwin, February 17, 2010; and John Reinhart, supervisor, Accessions and Control/Reference Sections, Illinois State Archives, e-mail message to Louis Gwin, February 18, 2010. Jesse Caleb was elected to the Twenty-Seventh Illinois General Assembly in

1870 and served one term in the Senate in 1871–72. According to his obituary, published in the Carthage *Republican* in 1917 and reproduced in Hart, he was appointed to a special Senate committee created to make recommendations about state compensation for the victims of the great Chicago fire of 1871. The Illinois State Historical Society documented Jesse Caleb's term in the Illinois General Assembly, but his service on the Senate committee to assist victims of the Chicago fire (or evidence that such a committee existed) could not be confirmed.

37. Marsha Lea Daggett, ed., *Pecos County History* (Canyon, Texas: Staked Plains Press, 1984), 2:517.

38. Hart, "Col. Richard Callaway," 2:491–509.

39. Michael Willis, *Billy the Kid: The Endless Ride* (New York: Norton, 2007), 242–44; and "Lew Wallace," *Wikipedia,* last modified August 16, 2010, http://en.wikipedia.org/

40. This synopsis of the early history of Fort Stockton, unless otherwise noted, is taken from Williams, *Texas' Last Frontier.*

41. Ibid, 273. The name was changed in the election of 1881 to honor both the historic old garrison and the man for whom the garrison was named, Robert Field Stockton, a United States Senator from New Jersey and a naval commodore in the Mexican-American War of 1846–48.

42. "Overview and History of Pecos County," Pecos County, last modified September 9, 2006, http://www.co.pecos.tx.us/. Pecos County was established by the Texas legislature in 1871 but was much larger than it is today. In 1883 Reeves and Terrell Counties were created from part of Pecos County, and in 1885 another part was incorporated into Val Verde County. Even so, Pecos County is the second largest county in Texas, encompassing 4,776 square miles.

43. Williams, *Texas' Last Frontier,* 325–26.

44. Daggett, *Pecos County History,* 1:109–12; and Williams, *Texas' Last Frontier,* 316–17. O. W.'s first large project involved surveying a tract of land southwest of Fort Stockton in 1885 that was three and one-half miles wide and thirty miles long, and today is shown on land maps as "Block O. W."

45. Clayton Williams, in recorded discussions with his family, transcript, undated, in the authors' possession; and "Christmases of Yesterday Recalled At Luncheon," *Fort Stockton Pioneer,* December 23 and 26, 1982. Clayton's family recorded interviews with their father several months before his death in 1983.

46. Williams, discussions with his family.

47. Williams, unpublished manuscript.

48. Williams, discussions with his family.

49. Daggett, *Pecos County History,* 1:131–35; and Williams, unpublished manuscript.

50. Williams, discussions with his family.

51. Ermine Williams, unpublished manuscript, no title or date, Fort Stockton Public Library, Fort Stockton, Texas.

52. Myres, ed., *Pioneer Surveyor, Frontier Lawyer,* 319n1. By 1913 there were 2,360 residents in Pecos County.

53. Daggett, *Pecos County History,* 1:246–47; and Harriet Herns to Janet Pollard, October 21, 2009, in the authors' possession. Jesse Williams Hart received her degree from Carthage College in 1883 and "taught for six terms," until 1889, when she went to work for a publishing house in Chicago. She married A. E. Hart in 1902 and moved to Los Angeles. Susan never married and devoted much of her life to caring for her parents in Illinois. She also cared for the Williams's children when they attended school in Illinois

and grew very close to her nieces and nephews. She eventually moved to Alpine, Texas, where she died in 1959.

54. Williams, *Texas' Last Frontier,* 520; and Williams, discussions with his family. Clayton said that his father sometimes walked the twenty-eight miles between the ranch and Fort Stockton.

55. Williams, discussions with his family.

56. Unless otherwise noted, the following summary of the feud is taken from Williams, *Texas' Last Frontier,* 339–88. Some of the facts concerning Deputy U.S. Marshall George Scarborough's involvement are taken from Robert K. DeArment, *George Scarborough: The Life and Death of a Lawman on the Closing Frontier* (Norman: University of Oklahoma Press, 1996), 53–57.

57. Clayton Williams's account of the feud is drawn from numerous notes and records that his father accumulated about the incident as well as from personal conversations between father and son. Janet Pollard recalls her mother's asking her father not to publish the story of the feud because it might stir up old resentments.

58. A building named The Gray Mule still stands at the corner of Main and Callaghan Streets in Fort Stockton. Koehler's store and saloon stands across Spring Drive from the Comanche Springs Pavilion and swimming pool; a second story was added in the 1930s. The building is now used as a small community hall.

59. Daggett, *Pecos County History,* 2:416.

60. Ibid., 2:415–18, 418–23.

61. Mary Alice Happle Townsend, "Andrew Jackson Royal," *Permian Historical Annual* 24 (1985): 51–60.

62. "To Mexico for Revenge: Sent to Brooklyn for Filibustering, Ochoa Tells His Story," *New York Times,* August 17, 1895, http://query.nytimes.com/.

63. George Morgenthaler, *The River Has Never Divided Us: A Border History of La Junta de los Rios* (Austin: University of Texas Press, 2004), 131–32.

64. O. W. W. to S. A. Purinton, November 23, 1894, in the authors' possession.

65. Ibid.

66. Daggett, *Pecos County History,* 2:420.

67. O. W. W. to Purinton; and Bill C. James and Mary Kay Shannon, *Sheriff A. J. Royal: Fort Stockton, Texas* (1984), 31.

68. Williams, *Texas' Last Frontier,* 392.

69. James and Shannon, 31.

Chapter 2

1. Williams, unpublished manuscript.

2. Williams, *Texas' Last Frontier,* 410.

3. Herns to Pollard.

4. Janet Pollard, Clay Pollard, and Scott Pollard, discussion with Louis Gwin, May 16, 2009.

5. Williams, discussions with his family; and "State Listings," National Register of Historic Places, accessed August 19, 2010, http://www.nationalregisterofhistoricplaces .com/. The only other remaining original building is the guardhouse. Other buildings have been reconstructed, and a museum is located on the site and open to the public for an admission fee. The fort was listed in the National Register of Historic Places in 1973.

6. Williams, discussions with his family.

7. Janet Pollard, discussion with Louis Gwin, June 30, 2008.

8. "Texas Historic Sites Access," Texas Historical Commission, accessed August 19, 2010, http://atlas.thc.state.tx.us/.

9. Ibid.; Daggett, *Pecos County History,* 2:522; and Myres, ed., *Pioneer Surveyor, Frontier Lawyer,* 319n5.

10. Also known as title plants, an abstract plant is used to issue title insurance policies in the State of Texas. It consists of indexed records showing all plat or map records, deeds, deeds of trust, mortgages, and other instruments of record affecting lands within a county for a period of at least twenty-five years immediately prior to the date of the title search. For a complete definition see "Title Plant," Integrity Title Records, accessed December 10, 2010, http://www.integritytitlerecords.com/home.html.

11. Maurice Bullock to C. W. W., July 30, 1964, in the authors' possession.

12. Williams, discussions with his family.

13. Ibid.

14. Ibid.

15. Ibid.; and "Horses Only Victims, Tornado Visits in 1903," *Fort Stockton Pioneer,* October 8, 1970.

16. Williams, discussions with his family.

17. "Rodeo Hall of Fame," National Cowboy & Western Heritage Museum, accessed August 19, 2010, http://www.nationalcowboymuseum.org/.

18. Williams, discussions with his family.

19. Ibid.

20. O. W. W. to Ermine Williams, n.d., in the authors' possession.

21. Janet Pollard and Scott Pollard, tape-recorded discussion with Clay Pollard, September 2008, transcript in the authors' possession.

22. Williams, discussions with his family.

23. Janet Pollard and Scott Pollard, discussion with Clay Pollard.

24. Ibid. "Caliche" is dirt encrusted with calcium carbonate; also called hardpan.

25. "Early Days and Incidents of FS Reviewed by Williams," *Fort Stockton Pioneer,* April 20, 1970.

26. Ibid. Donkey baseball is still played, often as a fund-raiser, although it is more common today as donkey basketball or "donkey ball" and is played in gymnasiums. For example, see "Donkey Ball Stubbornly Holds On despite Criticism," *New York Times,* April 18, 2009.

27. O. W. W. to J. C. W, June 14, 1925.

28. Williams, discussions with his family.

29. Janet Pollard, Clay Pollard, and Scott Pollard, discussion with Louis Gwin.

30. Daggett, *Pecos County History,* 1:203.

31. Williams, unpublished manuscript. A transcript of the January 3, 1912, letter from Clayton to his parents is included in the manuscript.

32. Ibid.

33. O. W. W. to J. C. W., June 16, 1918. J. C. Williams first attended college for a year at Texas A&M before transferring to New Mexico A&M.

34. Bobby Hawthorne, e-mail message to Louis Gwin, September 25, 2009.

35. Gaye Denley, "Early Aggie Footballer Reflects on Rich Past," *Battalion,* October 26, 1981; and Henry C. Dethloff, *A Pictorial History of Texas A&M University, 1876–1976* (College Station, Texas: Texas A&M University Press, 1975), 37.

36. Williams, unpublished manuscript.

37. Ibid.

38. Janet Pollard, Clay Pollard, and Scott Pollard, discussion with Louis Gwin.

39. Janet Pollard and Scott Pollard, discussion with Clay Pollard.

40. Ibid.

41. Janet Pollard, Clay Pollard, and Scott Pollard, discussion with Louis Gwin.

42. Ibid; and Janet Pollard and Scott Pollard, discussion with Clay Pollard.

43. Clayton's academic transcript from the Agricultural and Mechanical College of Texas, in the authors' possession; *Bulletin of the Agricultural and Mechanical College of Texas* 11, no. 4 (1925): 324; and *The Long Horn* (College Station: Texas Agricultural and Mechanical College, 1915), 78.

44. Williams, unpublished manuscript; "Commissioners," Railroad Commission of Texas, accessed August 19, 2010, http://www.rrc.state.tx.us/; "William D. Williams," *Handbook of Texas Online*, http://www.tshaonline.org/; and Sonnischen, intro., *Pioneer Surveyor, Frontier Lawyer*, 6–7.

45. C. W. W. to O. W. W, February 1, 1917, Clayton Wheat Williams papers, Haley Library; and "Chino," InfoMine, accessed August 19, 2010, http://www.infomine.com/. Hurley is about fourteen miles southeast of Silver City, New Mexico. The mine is now owned by Phelps Dodge.

46. Williams, undated, handwritten chronology, in the authors' possession.

47. C. W. W. to Sallie Wheat Williams and Ermine Williams, January 29, 1917, Haley Library.

48. Ibid.

49. Williams, unpublished manuscript.

50. Robert H. Zieger, *America's Great War: World War I and the American Experience* (Lanham, Maryland: Rowman & Littlefield, 2000), 22; and Ralph W. Steen, "World War I," *Handbook of Texas Online*, http://www.tshaonline.org/. On May 7, 1915, a German submarine torpedoed the British passenger liner *Lusitania* off the coast of Ireland, killing 1,200 people, including 128 U.S. citizens.

51. Friedrich Katz, *The Life and Times of Pancho Villa* (Stanford: Stanford University Press, 1998); Zieger, *America's Great War*, 50–51; and Steen, "World War I," *Handbook of Texas Online*, http://www.tshaonline.org/. PanchoVilla, born José Doroteo Arango Arámbula, was a popular Mexican revolutionary leader who formed his own military force in the northern Mexican state of Chihuahua and fought against the Mexican government from 1913 until the revolution ended in 1920.

Chapter 3

1. "Kroonland," Naval History and Heritage Command, accessed August 19, 2010, http://www.history.navy.mil/. The *Kroonland* was a passenger ship built in 1902 by W. Cramp Sons at Philadelphia for the International Mercantile Marine Company's Red Star line. In 1916 she was transferred to the New York-Liverpool run, and the navy placed guns and an armed naval guard on the ship to protect her from German submarines. On the morning of May 20, 1917, a torpedo struck her without exploding, and she collided with the German submarine that had fired it. After being repaired, she took up duty as a supply and troop transport for both the army and navy until the armistice. She was returned to her owner in 1919.

2. Clayton Williams, "World War I," unpublished manuscript [1983], typewritten, in the authors' possession.

3. Clayton Williams to My Dear Father, Serial No. 1 (revised), November 11, 1917, Haley Library. In addition to a number of shorter letters written during the war, Clayton

wrote thirty-two long letters that he labeled serials and numbered sequentially. At some point after the war, he typed and expanded the hand-written serials, providing a fuller description of his experiences while improving sentence structure and organization. The revisions also provided information that he could not include in the original letters due to censorship rules.

4. Williams, "World War I."

5. C. W. W. to Dear Father, Serial No. 2 (revised), n.d., Haley Library.

6. Ibid.

7. Williams, discussions with his family; and Williams, "World War I." The description of the Sam Browne belt is from Byron Farwell, *Over There: The United States in the Great War, 1917–1918* (Norton: New York, 1999), 304.

8. Williams, discussions with his family; and Williams, "World War I."

9. Zieger, *America's Great War*, 37–38.

10. Ibid., 58–86; and Farwell, *Over There*, 67.

11. Williams, discussions with his family.

12. Gus C. Dittmar, *They Were First: Recollections of the First Officers Training Camp of Leon Springs, Texas, May 8 to August 15, 1917* (Austin: Steck-Warlick, 1969), 4. Clayton attended reunions of First Campers throughout his life.

13. Williams, discussions with his family.

14. Dittmar, *They Were First*, 82.

15. Williams, discussions with his family.

16. Robert Arthur, *The Coast Artillery School: 1824–1927* (Fort Monroe, Virginia: Coast Artillery School Press, 1928), 61; and "About Fort Monroe," Fort Monroe, accessed August 19, 2010, http://www.monroe.army.mil/.

17. Arthur, *Coast Artillery School*, 63, 68.

18. Williams, discussions with his family.

19. C. W. W. to Dear Father, Serial No. 3 (revised), November 24, 1917, Haley Library. For an informative and well-researched account of the Halifax harbor disaster, see Laura M. Mac Donald, *Curse of the Narrows* (New York: Walker, 2005).

20. C. W. W. to Dear Folks, Serial No. 7 (revised), n.d., Haley Library; and Zieger, *America's Great War*, 113. In 1917, the year that Clayton sailed to Europe, Germany destroyed more than 6 million tons of shipping, nearly half of the total tonnage lost during the war. However, the use of American destroyers in the war zone and the adoption of the convoy system significantly reduced the number of sinkings.

21. C. W. W. to Dear Father, Serial No. 4 (revised), Haley Library.

22. C. W. W. to Dear Father, Serial No. 6 (revised), n.d., Haley Library.

23. C. W. W., Serial No. 7 (revised).

24. Ibid.

25. Williams, discussions with his family.

26. C. W. W., Serial No. 7 (revised).

27. C. W. W. to Dear Father, Serial No. 5 (revised), November 25, 1917, Haley Library; and Zieger, *America's Great War*, 113. This is possibly a description of the incident that occurred on May 20, 1917 (n1). By that year, the use of convoys and destroyers had begun to outweigh the initial German advantage in submarine warfare. The speed of the destroyers, improved depth charges, increased patrols and mining of the North Sea, and the inability of the Germans to replace destroyed submarines severely crippled the German submarine effort.

28. C. W. W. to Dear Father, Serial No. 9 (revised), December 2, 1917, Haley Library.

29. Ibid.

30. Zieger, *America's Great War*, 91.

31. Robert B. Bruce, *A Fraternity of Arms: America and France in the Great War* (Lawrence: University of Kansas Press, 2003), 97–143.

32. C. W. W., Serial No. 9 (revised).

33. C. W. W. to My Dear Folks, November 1, 1917, in the authors' possession.

34. C. W. W. to Dear Father, Serial No. 14 (revised), January 27, 1918, Haley Library.

35. Ian Hogg, *Allied Artillery of World War One* (Wiltshire, England: Crowood Press, 1998), 149–74; and John Batchelor and Ian Hogg, *Artillery* (New York: Charles Scribner's Sons, 1972), 64–68.

36. P. H. Ottosen, ed., *Trench Artillery A.E.F.: The Personal Experiences of Lieutenants and Captains of Artillery Who Served with Trench Mortars* (Boston: Lothrop, Lee & Shepard, 1931), 45; and C. W. W., Serial No. 14 (revised).

37. "Weaponry," firstworldwar.com, accessed August 19, 2010, http://www.firstworldwar.com/; Ottosen, *Trench Artillery*, 19–20; and Bruce, *Fraternity of Arms*, 104. By 1914, the Germans had built and were using very efficient mortars known as *minenwerfers* (mine-throwers), which they had developed after observing the Russo-Japanese War of 1904–05.

38. Ottosen, *Trench Artillery*, 21.

39. The Book of Job in the Old Testament describes Job as a righteous man whom God permits Satan to test with a series of misfortunes to see if Job will renounce him. Job does not complain or curse God during his ordeal.

40. Williams, "World War I."

41. C. W. W., Serial No. 14 (revised).

42. C. W. W. to Dear Folks, November 1, 1917, Haley Library; and C. W. W. to Dear Father, Serial No. 9 (A) (revised), October 1917, Haley Library.

43. C. W. W., Serial No. 9 (A) (revised); and Ottosen, *Trench Artillery*, 46.

44. Williams, "World War I." This story was not mentioned in the revised serial letters.

45. C. W. W., Serial No. 9 (A) (revised).

46. C. W. W. to Dear Father, Serial No. 12 (revised), December 9, 1917, Haley Library.

47. War Department, *Manual for Trench Artillery, United States Army (Provisional)*, Part I, *Trench Artillery* ([Washington, DC?]: Office of the Adjutant General, 1918; reprint, 2006), available online at http://cgsc.cdmhost.com/u?/p4013coi19,133; and "Weaponry," Firstworldwar.com, accessed August 19, 2010, http://www.firstworldwar.com/.

48. Ottosen, *Trench Artillery*, 15.

49. C. W. W., Serial No. 9 (A) (revised).

50. Williams, "World War I." This story was not mentioned in the revised serial letters.

51. C. W. W. to Dear Folks, Serial No. 11 (revised), December 9, 1917, Haley Library.

52. C. W. W. to Dear Father, Serial No. 10 (revised), December 8, 1917, Haley Library.

Chapter 4

1. Ottosen, *Trench Artillery*, 25.

2. C. W. W., Serial No. 14 (revised).

3. Ibid.; and Margaret H. Darrow, *French Women and the First World War: War Stories of*

the Home Front (Oxford: Berg, 2000), 79–89. The *marraines de guerre* were an outgrowth of Catholic women's charities in France that had attempted to convert prison inmates and conscripts through letter writing. When the Germans occupied northern France in 1915, the letter writing evolved into a wartime mission for French women to provide comfort and support for soldiers.

4. C. W. W., Serial No. 14 (revised); and Anne-Sophie Chevalier, genealogist, to Janet Pollard, October 13, 1998, in the authors' possession. Perhaps Suzanne Sykes-Gaubert chose American rather than French soldiers because she was married to an American. Janet and Bob Pollard visited France in 1997 to retrace her father's duty posts and travels during the First World War, and they enlisted the help of Chevalier to locate the Gaubert family. Through this contact, they learned that Suzanne Sykes-Gaubert had died in 1991, and her brother, some years earlier. One of Suzanne's daughters, Marguerite, told Janet Pollard in a telephone conversation after the trip that her family had loved Clayton.

5. C. W. W. to Dear Father, Serial No. 13 (revised), December 9, 1917, Haley Library.

6. Ibid.

7. Williams, discussions with his family. For information on the life of Raymond Poincaré, see J. F. V. Keiger, *Raymond Poincaré* (Cambridge: Cambridge University Press, 1997).

8. Janet Pollard, discussion with Louis Gwin, June 2007; O. W. W. to J. C. W., June 1, 1924, Haley Library; and Williams, discussions with his family.

9. C. W. W. to O. W. W., March 20, 1918, Haley Library.

10. C. W. W., Serial No. 14 (revised); and "Langres," *Classic Encyclopedia,* accessed August 20, 2010, http://www.1911encyclopedia.org/.

11. C. W. W. to Dear Father, n.d., Serial 14 (A) (revised), Haley Library; Ottosen, *Trench Artillery,* 25; "Rainbow Division Veterans Memorial Foundation, Inc.," Rainbow Division Veterans Memorial Foundation, accessed August 20, 2010, http://rainbowvets .org/; and C. W. W., Serial No. 14 (revised).

12. C. W. W. to Dear Father, December 25, 1917, Haley Library.

13. Ibid.

14. Myron Fox, "War Letters," *American Experience,* Public Broadcasting System, accessed August 20, 2010, http://www.pbs.org/wgbh/amex/.

15. C. W. W., Serial No. 10 (revised) Haley Library.

16. C. W. W. to "Dear Father," Serial No. 19 (revised), April 13, 1918, Haley Library.

17. C. W. W., Serial No. 14 (revised); and Ottosen, *Trench Artillery,* 43.

18. Graham Pollard, letter to Louis Gwin, n.d., in the authors' possession.

19. C. W. W., Serial No. 14 (revised); Kenneth E. Hamburger, *Learning Lessons in the American Expeditionary Forces* ([Washington, DC?]: U.S. Army Center of Military History, 1997), 15; and John Keegan, *The First World War* (New York: Vintage, 2000), 329–31.

20. C. W. W. to Dear Mother, February 4, 1918, Haley Library; and "Scabies," Center for Disease Control, accessed August 20, 2010, http://www.cdc.gov/. Scabies is a contagious infestation of the skin caused by the itch mite *Sarcoptes scabiei.*

21. C. W. W. to Dear Father, Serial 15 (A) (revised), n.d., Haley Library.

22. Ibid.; and Keegan, *First World War,* 356. In this letter, Clayton misspelled Steenvoorde as "Steeugroude." Bickebusch is no longer shown on the current Michelin map of France. The village of Messines and the Messines Ridge, located southeast of Ypres, was the site of a major British Army offensive in June 1917.

23. C. W. W., Serial 15 (A) (revised).

24. Keegan, *First World War,* 359; and C. DeF. Chandler, "Military Observation Balloons," *Field Artillery Journal* 7, no. 1 (1917): 15–20, http://sill-www.army.mil/FAMAG/archives.htm.

25. C. W. W., Serial No. 14 (revised).

26. Williams, discussions with his family. Clayton tells the same story in a revised serial, but the liquor in that retelling is identified as Scotch, not Cognac.

27. Ibid.; and Keegan, *First World War,* 421–23. Although considerable effort was expended on the Western Front to give the fallen a proper burial, by the war's end thousands of soldiers were never found or identified; by accident or expediency, they were often buried where they fell.

28. Clay Pollard, tape-recorded reminiscences, June 26, 2009, in the authors' possession.

29. Graham Pollard to Louis Gwin, n.d.

30. C. W. W., Serial No. 14 (revised).

31. C. W. W. to Dear Father, December 23, 1917, Haley Library; and Hart, "Col. Richard Callaway," 2:519, 528 (approximate page numbers; some pages in this volume were misnumbered as later material was added by Hart).

32. O. W. W. to J. C. W., July 23 and 25, 1918, Haley Library.

33. C. W. W. to Dear Father, December 23, 1917.

34. Ibid. Military historians have learned much after years of study about these two battles. See Keegan, *The First World War* and Tim Travers, *How the War Was Won: Command and Technology in the British Army on the Western Front, 1917–1918* (London: Routledge, 1992) for analysis of how the British top-down command system contributed to the loss of territory initially gained in the tank attack at Cambrai.

35. C. W. W., Serial No. 14 (revised).

36. Farwell, *Over There,* 99; Zieger, *America's Great War,* 91–92; Stephen Pope and Elizabeth-Anne Wheal, *The Dictionary of the First World War* (New York: St. Martin's, 1995), 492; John Ellis and Michael Cox, *The World War I Databook* (London: Aurum, 2001), 270; and Joseph F. Siler, *The Medical Department of the United States Army in the World War,* vol. 9, *Communicable and Other Diseases* (Washington, DC: U.S. Government Printing Office, 1928), 203–21.

37. C. W. W., Serial No. 15 (A); and Lavinia L. Dock et al., *History of American Red Cross Nursing* (New York: MacMillan, 1922), 493, http://books.google.com/.

38. Walter A. Payne, ed., *Benjamin Holt: The Story of the Caterpillar Tractor* (Stockton, California: University of the Pacific, 1982), 62–80.

39. W. R. Connolly, "Motor Transportation for Artillery," *Field Artillery Journal* 9, no. 3 (1919): 274.

40. C. W. W. to Dear Father, Serial No. 16 (revised), March 18, 1918; and Hogg, 94–96. *Grande puissance* is translated as "high power." The *Filloux* refers to Lt. Col. L. J. F. Filloux, who proposed the Canon de 155 GPF in 1916 as a long-range gun that could cover a wide target area. The gun went into serious production in 1917.

41. C. W. W., Serial No. 16 (revised).

42. C. W. W. to Dear Father, March 11, 1918.

43. C. W. W. to Dear Father, March 15 and March 28, 1918; and Dock et.al., 538–39. American Red Cross Military Hospital No. 3 was an officers' hospital established in 1915 in a building housing a club for girls at 4 rue de Chevreuse in Paris. It was taken over by the Red Cross in 1917. Clayton used the hospital's stationery, with letterhead and return address, to write some of his letters home.

44. C. W. W. to Dear Father, Serial No. 18 (revised), March 19, 1918, Haley Library; and John W. R. Taylor, *Jane's Fighting Aircraft of World War I* (London: Random House, 2001), 155–59. The Gotha Bomber was a twin-engine biplane built by the Gothaer Waggonfabrik A.G. Aircraft Company and introduced in 1916. It was extensively used by the Germans in both daylight and nighttime bombing missions over London and other targets in England starting in 1917.

45. Alice Ziska Synder and Milton Valentine Snyder, *Paris Days and London Nights* (New York: E. P. Dutton, 1921), 54–55, http://books.google.com/.

46. Snyder and Snyder, 55.

47. "Explosions Kill 16, Shake City of Paris," *New York Times,* March 15, 1918, http://query.nytimes.com/.

Chapter 5

1. "One Hundred Die in Paris from German Raid, 66 Suffocated," *New York Times,* March 13, 1918, http://query.nytimes.com/; "Sixty German Airplanes in Night Raid on Paris," *New York Times,* March 12, 1918, http://query.nytimes.com/; and Walter G. Green III, ed., *Electronic Encyclopaedia of Civil Defense and Emergency Management,* s.vv. "Strategic Bombers of World War I" and "Strategic Bombing in World War I—Germany," accessed August 22, 2010, http://www.richmond.edu/~wgreen/encyclopedia.htm.

2. C. W. W., Serial No. 18 (revised).

3. Ibid.

4. Ibid.

5. Ibid.; and Jennifer D. Keene, *Doughboys, the Great War, and the Remaking of America* (Baltimore: Johns Hopkins University Press, 2001), 118–19. Clayton's attitudes toward French merchants were shared by many American soldiers.

6. C. W. W., Serial No. 18 (revised).

7. C. W. W. to Dear People, Serial No. 20 (revised), April 15, 1918, Haley Library.

8. Ibid.

9. C. W. W., Serial No. 19 (revised).

10. Ibid.

11. Anne Cipriano Venzon, ed., *The United States in the First World War: An Encyclopedia* (New York: Garland, 1995), s.vv. "Brest Litovsk, Treaty of" and "Paris Gun."

12. C. W. W. to Dear Father, March 28, 1918; Venzon, *United States in the First World War,* s.v. "United States Air Service"; Keegan, *First World War,* 373; and Bruce, *Fraternity of Arms,* 197.

13. C. W. W. to Dear Sister, April 5, 1918, Haley Library; C. W. W., Serial No. 19 (revised); Pope and Wheal, 172; and Bruce, *Fraternity of Arms,* 204.

14. C. W. W., Serial No. 19 (revised); Keegan, *First World War,* 403–04; and Travers, *How the War Was Won,* 84–89. While Keegan credited Allied resistance as a factor in halting the German attack, he also believed that the offensive of March and April 1918 was stopped because the Germans diluted their attack along several locations rather than making a single thrust at the breakout point between the British and French armies. Also, German troops halted their advance to loot the British rear areas. Travers gave credit to British artillery fire for halting the German attack, though he blamed German infantry tactics (using large numbers of men to try to overwhelm British positions) as a contributing factor to the artillery's success.

15. C. W. W. to Dear Sister, April 5, 1918.

16. C. W. W., Serial No. 20 (revised).

17. Ibid.; and Frederick S. Mead, ed., *Harvard's Military Record in the World War* (Boston: Harvard Alumni Association, 1921), 343, http://books.google.com/.

18. C. W. W., Serial No. 20 (revised); and "The 56th Artillery, C.A.C.," rootsweb.ancestry.com, accessed August 22, 2010, http://freepages.military.rootsweb.ancestry.com/.

19. C. W. W., Serial No. 20 (revised).

20. Ibid.

21. Keene, 39.

22. C. W. W. to Dear Folks, Serial Nos. 21, 22, and 23, April 17, 1918, Haley Library; Serial No. 24, April 19, 1918, Haley Library; Serial No. 25, April 20, 1918, Haley Library; and Serial No. 26, n.d., Haley Library. Serial No. 26 contains the drawings by Clayton to illustrate points in the text.

23. C. W. W. to Dear Father, July 20, 1918, Haley Library.

24. C.W.W. to Dear Father, August 8, 1918, Serial No. 28 (revised), Haley Library.

25. Ibid.

26. C. W. W. to Dear Father, August 12, 1918, Haley Library; and Keegan, *First World War,* 410–11.

27. C. W. W. to Dear Father, August 8, 1918.

28. James D. Crabtree, *On Air Defense* (Westport, Connecticut: Praeger, 1994), 35, http://books.google.com/.

29. C. W. W. to Dear Father, May 9, 1918, Haley Library; and ibid., Serial 28 (A) (revised), July 20, 1918, Haley Library.

30. C.W.W. to Dear Father, May 9, 1918; and Keegan, *First World War,* 408.

31. C. W. W. to O. W. W., June 9, 1918, Haley Library.

32. C. W. W. to O. W. W., July 20, 1918.

33. Williams, unpublished manuscript.

34. C. W. W. to O. W. W., June 7, 1918, Haley Library.

35. C. W. W. to O. W. W., June 23, 1918, Haley Library; and "Victor Herbert," *Wikipedia,* accessed August 22, 2010, http://en.wikipedia.org/. Victor Herbert was one of America's finest cellists from the 1880s well into the 1910s. He also composed hundreds of popular songs and marches and major classical works for cello and orchestra, as well as numerous symphonic works, operettas, and movie scores, including one of the first fully original American film scores, *Fall of a Nation.* He is perhaps best known as the composer of the operetta *Babes in Toyland.* He died in 1924 at the age of sixty-five.

36. C. W. W., Serial No. 28 (A) (revised). Clayton said he trained nearly 300 men in auto mechanics in two courses at the center.

37. C. W. W. to Dear Father, December 10, 1918, Haley Library.

38. C. W. W. to Dear Father, December 12, 1918, Haley Library; and John F. Votaw, *The American Expeditionary Forces in World War I* (Oxford: Osprey, 2005), 82.

39. Clayton, unpublished manuscript; and Special Orders No. 53, Headquarters, O &T Center, TA, #3, AEF, July 18, 1918, in the authors' possession.

40. Keegan, *First World War,* 411.

41. C. W. W. to Dear Father, October 13, 1918, Haley Library.

42. Ibid.

43. Ibid.

44. "One Hundred Nineteen Casualties in Army and Marine Corps," *New York Times,*

July 1, 1918, http://query.nytimes.com/. Clayton frequently used only last names in referring to friends and associates in his letters. Thanks to Clay Pollard, an A&M graduate,
for locating the first names of Gilfillan and Brailsford.

45. Clayton, unpublished manuscript; and C. W. W. to O. W. W., January 18, 1919,
Haley Library.

46. C. W. W. to Dear Father, December 10, 1918.

47. Ibid; P. C. Harris, Adjutant General, Memorandum of Information, March 25, 1919,
in the authors' possession; and C. W. W. to P. C. Harris, Adjutant General, April 21, 1919,
in the authors' possession.

48. C. W. W. to Dear Father, December 12, 1918.

49. Payne, *Benjamin Holt,* 25–37.

50. C. W. W. to Dear Father, December 12, 1918.

51. O. W. W. to J. C. W., November 17, 1918, Haley Library.

52. C. W. W. to O. W. W., December 22, 1918, Haley Library.

53. Ibid.

54. Philip S. Dickey III, "The Liberty Engine, 1918–1942," *Smithsonian Annals of Flight*
1, no. 3 (Washington: Smithsonian Institution Press, 1968); J. G. G. Hempson, "The Automobile Engine, 1920–1950," in *A History of the Automotive Internal Combustion Engine* (Warrendale, Pennsylvania: Society of Automotive Engineers, 1976), 13; and Votaw,
American Expeditionary Forces, 83.

55. C. W. W. to Dear Father, December 22, 1918.

56. C. W. W. to Dear Father, January 4, 1919, Haley Library.

57. C. W. W. to Dear Father, postmarked February 23, 1919, Haley Library.

58. Ibid.

59. C. W. W. to Dear Father, February 23, 1919, later labeled Serial No. 32 (A), Haley Library. The letter was written on the stationery of the Hotel Colonnade of Philadelphia.

Chapter 6

1. Ronald Allen Goldberg, *America in the Twenties* (Syracuse, New York: Syracuse University Press, 2003), 2–5; and Geoffrey Perrett, *America in the Twenties: A History* (New
York: Simon and Schuster, 1982), 29–30.

2. C. W. W. to O. W. W., February 27, 1919, Haley Library.

3. John Moody, *Moody's Analyses of Investments,* Part II, *Industrial Investments,* 1920
(New York: Moody's Investors Service, 1920), 448–51. Any references to 2009 dollars,
the latest year available, are calculated using the Consumer Price Index. We used the
calculator at the Measuringworth website, last modified April 2010, http://www.measur
ingworth.com/uscompare/.

4. C. W. W. to O. W. W., March 9, 1919, Haley Library.

5. C. W. W. to O. W. W., February 28, 1919; "Caruso Marries Miss Benjamin," *New
York Times,* August 21, 1918, http://query.nytimes.com/; "Metropolitan Announces Op-
eras," *New York Times,* February 10, 1919; and Niall Palmer, *The Twenties in America:
Politics and History* (Edinburgh: Edinburgh University Press, 2006), 7. Clayton does not
mention having alcohol at the dinner, perhaps because the Eighteenth Amendment to
the Constitution, outlawing liquor sales, had come into effect one month earlier, or because he did not wish to offend his father, who was a teetotaler.

6. C. W. W. to O. W. W., March 9, 1919.

7. "War Revenue Bill Perfected Is Put before Congress," *New York Times,* Febru-

ary 7, 1919, http://query.nytimes.com/; Dixon Wecter, *When Johnny Comes Marching Home* (Cambridge: Riverside, 1944), 312, 375–83; and Francis B. Eastman, "Valuable Information for Discharged Soldiers of the United States Army (1919)," Gjenvick-Gjønvik Archives, accessed August 23, 2010, http://www.gjenvick.com/Military/WorldWarOne/ Brochures/.

8. C. W. W. to O. W. W., March 10, 1919, Haley Library.

9. Williams, discussions with his family; and O. W. W. to J. C. W., April 12, 1919, Haley Library.

10. O. W. W. to J. C. W., April 27 and May 14, 1919, Haley Library.

11. Samuel D. Myres, *Era of Discovery, from the Beginning to the Depression,* vol. 1 of *The Permian Basin: Petroleum Empire of the Southwest* (El Paso: Permian Press, 1973), 54–61; and "Past Year Was Busiest in the History of Petroleum," *Oil and Gas Journal* 18, no. 34 (1920): 54.

12. Myres, *Era of Discovery,* 59. For example, in one year after the strike at Ranger, the town's population increased from 1,000 to 25,000.

13. Williams, discussions with his family.

14. O. W. W. to J. C. W., October 28, 1919, Haley Library.

15. O. W. W. to J. C. W., November 26, 1919, Haley Library; and Williams, unpublished manuscript.

16. O. W. W. to J. C. W., November 8 and November 18, 1919, Haley Library.

17. Myres, *Era of Discovery,* 113–17.

18. Williams, unpublished manuscript.

19. Myres, *Era of Discovery,* 117–18; and Williams, discussions with his family.

20. Myres, *Era of Discovery,* 119.

21. Ibid., 120.

22. R. D. Langenkamp, *Handbook of Oil Industry Terms & Phrases,* 4th ed. (Tulsa: PennWell, 1984). A swage nipple is a short length of pipe with threads or weld ends on both ends. The gate is a gate valve, a type of pipeline valve. The casing is steel pipe used to seal off fluids from the bore hole and prevent the walls of the hole from sloughing off or caving.

23. Williams, discussions with his family.

24. Charles C. Coulter, "Petroleum 'Graveyard Area' of Texas," *Oil and Gas Journal* 17, no. 48 (1919): 66–67; and Williams, discussions with his family.

25. Williams, unpublished manuscript.

26. Ibid.

27. Williams, discussions with his family.

28. O. W. W. to J. C. W., October 22 and November 19, 1922, Haley Library; and Williams, unpublished manuscript.

29. O. W. W. to J. C. W., November 30, 1922, Haley Library.

30. Williams, unpublished manuscript.

31. O. W. W. to J. C. W., April 13, 1924, Haley Library.

32. "Waldo Williams, Engineer, Ranchman and Naturalist, Buried Here Sunday Afternoon," *Fort Stockton Pioneer,* August 16, 1946.

33. Mari Helen Williams Schultz, *Waldo,* n.d., 1, in the authors' possession.

34. David Walker, e-mail message to Louis Gwin, August 10, 2009.

35. Martin Caidin, *Barnstorming* (New York: Duell, Sloan and Pearce, 1965).

36. O. W. W. to C. W. W., April 6, 1930, Haley Library.

37. Clayton W. Williams Jr., discussion with Louis Gwin, July 1, 2008, Midland, Texas.

38. "Carpenter Denies Foul," *New York Times,* August 14, 1908, http://query.nytimes .com/; Schultz, *Waldo,* 3; and O. W. W. to J. C. W., May 16, 1937, Haley Library.

39. O. W. W. to J. C. W., April 6, 1930; and "Réne Paul Fonck,"The Great War Flying Museum, accessed August 23, 2010, http://www.greatwarflyingmuseum.com/.

40. "Waldo Williams," *Fort Stockton Pioneer.*

41. O. W. W. to J. C. W., April 27, 1924, Haley Library.

42. O. W. W. to J. C. W., May 11, 1924, Haley Library.

43. O. W. W. to J. C. W., July 20, 1924, Haley Library.

44. O. W. W. to J. C. W., August 3, 1924, Haley Library.

45. O. W. W. to J. C. W., September 28 and October 12, 1924, Haley Library.

46. "Agreement," November 1, 1924, in the authors' possession; and Susan R. Walsh Sanderson, *Land Reform in Mexico: 1910–1980* (Orlando: Academic Press, 1984).

47. O. W. W. to J. C. W., December 21, 1924, Haley Library.

48. Schultz, *Waldo,* 12.

49. O. W. W. to J. C. W., December 21, 1924.

50. David Walker, e-mail message to Louis Gwin, August 10, 2009.

51. Myres, *Era of Discovery,* 205.

52. Ibid., 1.

53. Ibid., 194; David F. Prindle, "Oil and the Permanent University Fund: The Early Years," *Southwestern Historical Quarterly* 86, no. 2 (October 1982), 277–98; Martin W. Schwettmann, *Santa Rita* (Austin: Texas State Historical Association, 1943), 39–40; and Julia Cauble Smith, "Santa Rita Oil Well," *Handbook of Texas Online,* http://www.tsha online.org/.

54. Myres, *Era of Discovery,* 194–273.

55. Williams, discussions with his family.

56. Myres, *Era of Discovery,* 224–29.

Chapter 7

1. C. W. W. to Martin W. Schwettmann, May 16, 1940, in the authors' possession.

2. Williams, discussions with his family.

3. O. W. W. to J. C. W., March 7, 1926, Haley Library.

4. Janet Pollard and Scott Pollard, discussion with Clay Pollard.

5. C. W. W. to Scott Pollard, February 26, 1982, in the authors' possession; and Williams, unpublished manuscript.

6. Janet Pollard and Scott Pollard, discussion with Clay Pollard.

7. Adam Pollard, e-mail message to Louis Gwin, September 7, 2009.

8. Adam Pollard, e-mail message to Louis Gwin, November 4, 2009.

9. Myres, *Era of Discovery,* 267. S. O. Cooper represented Lee C. Moore and Company, which manufactured and sold most of the steel derricks used in Big Lake and other West Texas fields.

10. C. W. W. to Schwettmann.

11. Ibid.

12. Williams, discussions with his family; and Bryan B. Sterling and Frances N. Sterling, *Will Rogers & Wiley Post: Death at Barrow* (New York: M. Evans, 1993). Post set records for flying around the world in 1931 and 1933, and also developed the first pressure

suit for high altitude flying. Post and humorist Will Rogers were killed in an airplane crash in 1935 at Point Barrow, Alaska. Post was thirty-seven years old.

13. Williams, unpublished manuscript.

14. Ibid. In this version of the story, he identifies the lease broker as Sam Murray of Sheffield, Texas.

15. C. W. W. to Scott Pollard.

16. Ibid.; and O. W. W. to J. C. W., March 13, 1927, Haley Library.

17. Williams, unpublished manuscript.

18. Williams, discussions with his family.

19. Ibid.

20. Ibid.; Clayton W. Williams Jr., discussion with Louis Gwin; and George P. Hartmann, P.E., licensing project manager, Texas Board of Professional Engineers, e-mail message to Louis Gwin, September 21, 2009.

21. Williams, unpublished manuscript.

22. Ibid; and Robert O. Anderson, *Fundamentals of Petroleum Geology* (Norman: University of Oklahoma Press), 106. Reference books read by Clayton were Dorsey Hager, *Practical Oil Geology* (New York: McGraw Hill, 1916); and G. H. Cox, C. L. Dake and G. A. Muilenburg, *Field Methods in Petroleum Geology* (New York: McGraw Hill, 1921). The definition of what constitutes the "Big Lime" formation comes from Edgar Wesley Owen, *Trek of the Oil Finders: A History of Exploration for Petroleum* (Tulsa: American Association of Petroleum Geologists, 1975), 886.

23. Williams, discussions with his family. For an interesting description of the efforts of early geologists to find oil, see Anderson, *Fundamentals of Petroleum Geology,* 94–103.

24. C. W. W. to O. W. W., February 25, 1925, Haley Library.

25. Anderson, *Fundamentals of Petroleum Geology,* 91–92.

26. Owen, *Trek of the Oil Finders,* 489–91; and Anderson, *Fundamentals of Petroleum Geology,* 120–43.

27. Clarence Pope, *An Oil Scout in the Permian Basin: 1924–1960* (El Paso: Permian Press, 1972), 23–24.

28. O. W. W. to J. C. W., January 9, 1926, Haley Library.

29. Langenkamp, *Handbook,* 312.

30. Anderson, *Fundamentals of Petroleum Geology,* 101.

31. O. W. W. to J. C. W., February 15, 1925, Haley Library; Langenkamp, *Handbook,* 168; and Anderson, *Fundamentals of Petroleum Geology,* 97.

32. O. W. W. to J. C. W., January 17, 1926, Haley Library.

33. O. W. W. to J. C. W., November 7, 1926, Haley Library; and Railroad Commission of Texas, *Oil and Gas Division Annual Report, 2006,* vol. 1 (Austin: Railroad Commission of Texas, Oil and Gas Division, 2007), B-308.

34. Williams, discussions with his family.

35. O. W. W. to J. C. W., July 23, 1933, Haley Library.

36. O. W. W. to J. C. W., March 27, 1927, Haley Library.

37. O. W. W. to J. C. W., September 4, 1927, Haley Library.

38. O. W. W. to J. C. W., March 14, 1926. The reference to Gilpin is from a comic poem by the English poet and hymnodist William Cowper (1731–1800) that describes how a wealthy English draper was carried away on a runaway horse. See "The Diverting History of John Gilpin," Fullbooks.com, accessed August 24, 2010, http://www.full books.com/.

39. O. W. W. to J. C. W., February 15, 1925; Myres, *Era of Discovery*, 232; and "History," Natural Gas Supply Association, accessed August 24, 2010, http://www.naturalgas.org. Much of the natural gas discovered by drilling during this period was wasted. While many cities used natural gas to light street lamps for much of the 1800s, most of the gas was produced from coal. It took the development of reliable pipelines in the 1940s for natural gas from wells to reach new markets.

40. O. W. W. to J. C. W., January 24, 1926; Langenkamp, *Handbook*, 43; and Anderson, *Fundamentals of Petroleum Geology*, 238.

41. O. W. W. to J. C. W., February 7, 1926, Haley Library; and Anderson, *Fundamentals of Petroleum Geology*, 163–64. A technique used today to produce a similar result is called a "frac job." Water containing gels of various types and other fluids that work well with the particular formation being fractured is pumped into the hole at very high pressure to break up the producing formation around the well bottom. Sand or other material is then injected to keep the fractures open.

42. Myres, *Era of Discovery*, 249.

Chapter 8

1. Julia Cauble Smith, "Big Lake Oilfield," *Handbook of Texas Online*, http://www.tsha online.org/.

2. Myres, *Era of Discovery*, 230.

3. C. W. W. to Schwettmann.

4. "The Geologic Time Scale," United States Geological Survey, accessed August 24, 2010, http://vulcan.wr.usgs.gov/.

5. C. W. W. to Schwettmann; and Myres, *Era of Discovery*, 251.

6. Owen, *Trek of the Oil Finders*, 489–91; and Anderson, *Fundamentals of Petroleum Geology*, 120–43.

7. C. W. W. to Schwettmann; and Williams, discussions with his family.

8. C. W. W. to Schwettmann; and Myres, *Era of Discovery*, 251.

9. Langenkamp, *Handbook*, 100.

10. Myres, *Era of Discovery*, 250–54; and C. W. W. to Schwettmann. Myres said that drilling was halted after the second fire until June 11, but both Clayton and Waldo said that drilling was halted until August 16. See Waldo Williams, "Addends to Clayton Williams' Deepest Producing Oil Well in the World," December 1928, in the authors' possession.

11. "Waldo Williams," *Fort Stockton Pioneer*.

12. Waldo Williams, "Addends."

13. Ibid.; and Langenkamp, *Handbook*, 16.

14. Waldo Williams, "Addends."

15. Myres, *Era of Discovery*, 253.

16. Samuel D. Myres, *Era of Advancement, from the Depression to the Present*, vol. 2 of *The Permian Basin: Petroleum Empire of the Southwest* (El Paso: Permian Press, 1977), 3; and Owen, *Treck of the Oil Finders*, 904.

17. Janet Pollard, discussion with Louis Gwin, September 18, 2009.

18. O. W. W. to J. C. W., December 23, 1923, and December 21, 1924, Haley Library.

19. O. W. W. to J. C. W., April 4, 1926, Haley Library.

20. O. W. W. to J. C. W., March 7, 1926. J. C. was employed by the Texas Company (Texaco) in 1923 and left for China in March of that year to work as an auditor. See O. W. W. to J. C. W., April 12, 1923, Haley Library. By the time J. C. went to work for the

Texas Company, it had established marketing organizations for petroleum and petroleum products in western Europe, India, South Africa, Australia, Brazil, New Zealand, Cuba, Haiti, Japan, Puerto Rico, and China. See *Moody's Analyses of Investments and Security Rating Books, Industrial Investments* (New York: Moody's Investors Service, 1923), 833–35.

21. O. W. W. to J. C. W., August 1, 1926, Haley Library.

22. John Paul Pitts, "Profile of a Pioneer Wife," *Midland Reporter-Telegram,* October 25, 1987. Chicora Graham Williams said her unusual first name has its roots in the Civil War. "My grandfather, Captain Joe Graham of the Confederate Army, was a prisoner of war in a Yankee prison. The only way he survived was by the good graces of a little Indian girl, Chicora, who used to slip him bits of food." She said that Captain Graham vowed to name his first daughter for the Indian girl if he made it home. "My aunt was the first Chicora. I am the second Chicora and Claytie's daughter (Chim) is the third Chicora."

23. Chicora Graham Williams, untitled reminiscence, typewritten, June 14, 1984, in the authors' possession; and Williams, discussions with his family.

24. Chicora Williams, reminiscence; and C. W. W. to O. W. W., August 22, 1927, Haley Library.

25. Williams, unpublished manuscript.

26. Chicora Williams, reminiscence.

27. C. W. W. to Chicora Graham, August 29, 1927, in the authors' possession.

28. C. W. W. to Chicora Graham, October 19, 1927, in the authors' possession.

29. C. W. W to Chicora Graham, November 3, 1927, in the authors' possession.

30. C. W. W. to Chicora Graham, November 29, 1927, in the authors' possession.

31. C. W. W. to Chicora Graham, December 14, 1927, in the authors' possession.

32. Williams, unpublished manuscript.

33. C. W. W. to Chicora Graham, January 24, 1928, in the authors' possession.

34. C. W. W. to Chicora Graham, February 4, 1928, in the authors' possession.

35. C. W. W. to Chicora Graham, February 19, 1928, in the authors' possession; and J. Evetts Haley, Jim Trott, Berte R. Haigh, and Ford Chapman to Hall of Fame Committee, Permian Basin Petroleum Museum, Library and Hall of Fame, with attached "Biographical Information Clayton Wheat Williams Sr.," July 25, 1984, in the authors' possession.

36. Williams, unpublished manuscript.

37. C. W. W. to Chicora Graham, March 16, 1928, in the authors' possession.

38. C. W. W. to Chicora Graham, May 8, 1928, in the authors' possession.

39. C.W.W. to Chicora Graham, May 15, 1928, in the authors' possession.

Chapter 9

1. Williams, unpublished manuscript.

2. Ibid.; Myres, *Era of Discovery,* 498; and O. W. W. to J. C. W., May 2, 1929.

3. For an informative discussion of the history of proration, see Roger M. Olien and Diana Davids Hinton, *Wildcatters: Texas Independent Oilmen* (College Station: Texas A&M University Press, 2007), 43–66. Engineers theorized that heavy production from oil reservoirs lowered the pressure in the reservoirs and increased water incursion in the wells, although there was no technology then available to test the theory.

4. O. W. W. to J. C. W., July 17, 1929, Haley Library; O. W. W. to J. C. W., August 25, 1929; and Williams, discussions with his family.

5. C. W. W. to Myrtle Corley, February 21, 1973, in the authors' possession; and Roger Thévenot, *A History of Refrigeration throughout the World*, trans. J. C. Fidler (Paris: International Institute of Refrigeration, 1973), 172, 233.

6. Williams, discussions with his family.

7. Janet Pollard and Scott Pollard, discussion with Clay Pollard; Janet Pollard, Clay Pollard, and Scott Pollard, discussion with Louis Gwin.

8. O. W. W. to J. C. W., February 21 and April 13, 1929, Haley Library.

9. Clay Pollard, e-mail message to Louis Gwin, September 28, 2009.

10. O. W. W. to J. C. W., April 13 and September 1, 1929, Haley Library.

11. Myres, *Era of Advancement*, 152–53; O. W. W. to J. C. W., July 13, 1930, Haley Library; and Langenkamp, *Handbook*, 133.

12. Clayton W. Williams Jr., discussion with Louis Gwin.

13. O. W. W. to J. C. W., January 7 and January 19, 1930, Haley Library; and Williams, discussions with his family.

14. Olien and Hinton, 41–42.

15. Williams, discussions with his family; and Myres, *Era of Advancement*, 162.

16. H. Allen Anderson, "Thornton, Ward A. [Tex]," *Handbook of Texas Online*, http://www.tshaonline.org/. Thornton (1891–1949), who learned how to handle explosives while working for a company that manufactured torpedoes, established a company to manufacture nitroglycerin outside Amarillo and became known as the "king" of the oil-well fire fighters in the 1920s and 1930s. Ironically, Thornton was not killed doing his dangerous work but was murdered by two hitchhikers to whom he had given a ride.

17. Langenkamp, *Handbook*, 310. A tool dresser at a cable-drilling rig restores a sharp cutting edge to the drill bit by heating the bit and hammering the cutting edge until it is sharp.

18. O. W. W. to J. C. W., August 18, 1929, Haley Library.

19. O. W. W. to J. C. W., July 6, 1930, Haley Library.

20. O. W. W. to J. C. W., July 22, 1930, Haley Library.

21. O. W. W. to J. C. W., April 14 and 28, 1931, Haley Library.

22. Erika Murr, "Earthquakes," *Handbook of Texas Online*, http://www.tshaonline.org/.

23. O. W. W. to J. C. W., August 16, 1931, Haley Library.

24. O. W. W. to J. C. W., March 1, 1931, Haley Library.

25. Olien and Hinton, 56–58; and "Oil Prices Reduced by Humble in Texas," *New York Times*, October 17, 1930, http://query.nytimes.com/.

26. O. W. W. to J. C. W., April 1, 1931, Haley Library.

27. O. W. W. to J. C. W., April 28, 1931, Haley Library.

28. O. W. W. to J. C. W., January 23, 1932, Haley Library.

29. O. W. W. to J. C. W., July 12 and 19, 1931, Haley Library.

30. O. W. W. to J. C. W., August 9, 1931, Haley Library.

31. Williams, discussions with his family.

32. "First National Bank Failed to Open Tuesday," *Fort Stockton Pioneer*, October 9, 1931; and O. W. W. to J. C. W., November 1, 1931, Haley Library.

33. O. W. W. to J. C. W., September 27, 1931, Haley Library; and Williams, unpublished manuscript.

34. O. W. W. to J. C. W., January 23, 1932. A search of legal data bases and a request to the Rusk County District Court produced no records of the case. According to a

reference librarian at the Tarlton Law Library of the University of Texas School of Law, old Texas district court records are generally not kept by the court after a period of several years.

35. Ibid.; and O. W. W. to J. C. W., August 9, 1932, Haley Library.

36. Clayton W. Williams Jr., discussion with Louis Gwin.

37. O. W. W. to J. C. W., December 25, 1931, Haley Library.

38. Chicora Graham Williams to her children, June 14, 1984, in the authors' possession.

39. Pitts, "Profile of a Pioneer Wife."

40. Chicora Graham Williams to her children.

41. O. W. W. to J. C. W., December 26, 1932, Haley Library.

42. O. W. W. to J. C. W., December 3, 1932, Haley Library.

43. John D. Huddleston,"Ferguson, Miriam Amanda Wallace [Ma]," *Handbook of Texas Online,* http://www.tshaonline.org/.

44. Williams, unpublished manuscript.

45. O. W. W. to J. C. W., January 8, 1933, Haley Library.

46. O. W. W. to C. W. W., April 2, 1933, Haley Library.

47. O. W. W. to C. W. W., April 9,1933, Haley Library; and Williams, unpublished manuscript.

48. O. W. W. to J. C. W., March 19, 1933, Haley Library.

49. O. W. W. to J. C. W., April 9, 1933.

50. O. W. W. to J. C. W., April 2, 1933.

51. O. W. W. to J. C. W., February 12, 1933, Haley Library; and Olien and Hinton, 69.

52. O. W. W. to J. C. W., November 20, 1932; January 1, 1933; June 4 and 29, 1933; and July 9, 1933, Haley Library.

53. O. W. W. to J. C. W., July 16, 1933, Haley Library; *San Angelo (Texas) Morning Times,* August 2, 1933; and James Butkiewicz,"Reconstruction Finance Corporation," entry in online Economic History Encyclopedia, Economic History Association, accessed August 25, 2010, http://eh.net/.

54. O. W. W. to J. C. W., July 23, 1933, Haley Library.

55. O. W. W. to J. C. W., August 4 and September 9, 1933, Haley Library.

56. O. W. W. to J. C. W., February 17, 1934, Haley Library.

Chapter 10

1. "About Counties," Texas Association of Counties and V. G. Young Institute of County Government, accessed August 25, 2010, http://www.texascounties4u.org/.

2. Williams, discussions with his family. After Clayton took office in 1934, the commission hired someone for the position for $12,000 a year, which was $10,000 less than the old commission had paid. See also Daggett, *Pecos County History,* 1:115. A position titled "Assessor for Oil Properties" was filled by the commission in 1934.

3. "Clayton Williams Out for County Commissioner," *Fort Stockton Pioneer,* February 9, 1934.

4. A search of election records held in the Pecos County Court Clerk's office in Fort Stockton, Texas, found the results of Clayton's general elections but not of the primaries. Primary results were reported in the *Fort Stockton Pioneer,* but the issues covering the 1935 county Democratic primary are missing.

5. "Commissioner's Race: Attention Voters of Precinct No. 1," *Fort Stockton Pioneer,*

July 31, 1936; and "Some Reasons Why the Voters of Precinct Number One Should Re-Elect Clayton Williams, Commissioner," *Fort Stockton Pioneer*, August 14, 1936.

6. O. W. W. to J. C. W., August 9, 1933, Haley Library.

7. "Williams Files Suit Contesting Election of Cliett as Nominee," *Fort Stockton Pioneer*, September 11, 1936.

8. "Attention! To the Voters of Precinct No. One," *Fort Stockton Pioneer*, August 14, 1942; and "To The Voters," *Fort Stockton Pioneer*, August 21, 1943.

9. Williams, discussions with his family; and Janet Pollard, discussion with Louis Gwin.

10. Myres, *Era of Advancement*, 15–17.

11. O. W. W. to J. C. W., October 22, 1935, Haley Library; and Owen, *Trek of the Oil Finders*, 911.

12. O. W. W. to J. C. W., April 13 and 18, 1929; July 4 and 17, 1929; August 11, 1929; and July 13, 1929, Haley Library.

13. O. W. W. to J. C. W., April 6, 1934, Haley Library.

14. O. W. W. to J. C. W., April 22, 1934, Haley Library.

15. O. W. W. to J. C. W., September 8, 1935, Haley Library.

16. O. W. W. to J. C. W., September 15, 1935, Haley Library.

17. O. W. W. to J. C. W., January 12, 1936, Haley Library.

18. O. W. W. to J. C. W., April 26, 1936, Haley Library.

19. O. W. W. to J. C. W., May 17 and 30, 1936, Haley Library.

20. O. W. W. to J. C. W., August 9, 1936, Haley Library.

21. O. W. W. to J. C. W., July 4, 1936, Haley Library.

22. O. W. W. to J. C. W., November 15, 1936, Haley Library.

23. O. W. W. to J. C. W., February 28, 1937, Haley Library.

24. O. W. W. to J. C. W., March 14, 20, and 28, 1937; April 4, 18, and 25, 1937; May 16, 1937; and June 21, 1937, Haley Library.

25. O. W. W. to J. C. W., June 21, 1937.

26. O. W. W. to J. C. W., February 17, 1935, Haley Library.

27. O. W. W. to J. C. W., February 16, 1936, Haley Library.

28. O. W. W. to J. C. W., May 30, 1936.

29. O. W. W. to J. C. W., July 14, 1935, Haley Library; and Delmar J. Hayter, "Red Bluff Dam and Reservoir," *Handbook of Texas Online*, http://www.tshaonline.org/.

30. O. W. W. to J. C. W., August 25, 1935, Haley Library; K. Austin Kerr, "Prohibition," *Handbook of Texas Online*, http://www.tshaonline.org/; O. W. W. to J. C. W., September 22, 1935, Haley Library; and Cochran, *Claytie*, 20. In his biography, Claytie told a story about his father's concession to his grandfather's hatred of whiskey. Clayton Sr. was entertaining visitors with alcoholic drinks at his home in Fort Stockton when he spotted O. W. walking toward the house. Clayton immediately grabbed up the whiskey bottles and glasses and poured everything down the sink.

31. O. W. W. to J. C. W., August 30, 1936, Haley Library.

32. J. C. W. to C. W. W., October 3, 1936, Haley Library.

33. O. W. W. to J. C. W., November 22, 1936, Haley Library.

34. O. W. W. to J. C. W., September 20 and 27 and November 29, 1936, Haley Library.

35. Ermine Williams to J. C. W., undated, included in O. W. W. to J. C. W., November 22, 1936, Haley Library.

36. O. W. W. to J. C. W., November 26, 1936, Haley Library.

37. O. W. W. to J. C. W., November 15, 1936, Haley Library.

38. O. W. W. to J. C. W., June 21, 1937.

39. O. W. W. to J. C. W., January 31, 1937, Haley Library.

40. O. W. W. to J. C. W., February 28, 1937; March 14, 1937; and April 25, 1937, Haley Library.

41. Williams, discussions with his family.

42. Ibid.

43. Janet Pollard, discussion with Louis Gwin, December 10, 2010.

44. Ibid., June 30, 2008.

45. Chicora Graham Williams, in Williams, discussions with his family.

46. Williams, discussions with his family; and Robert Hotz, ed., *Way of a Fighter: The Memoirs of Claire Lee Chennault* (New York: C. P. Putnam's Sons, 1949), 232.

47. O. W. W. to J. C. W., October 8, 1942, Haley Library; Myres, *Era of Advancement,* 27; and Olien and Hinton, *Wildcatters,* 219–37.

48. Glenn Justice, "Gibbs Field," *Handbook of Texas Online,* http://www.tshaonline.org/; and O. W. W. to J. C. W., May 2 and July 27, 1942, Haley Library.

49. Cochran, *Claytie,* 41–42, 116; and Janet Pollard, conversation with Louis Gwin, September 11, 2010. Clayton's effort to raise goats is also mentioned in O. W. W. to J. C. W., October 8, 1942, Haley Library. Clayton briefly discusses Brangus cattle in Williams, discussions with his family. Claytie was at one time one of the leading purebred Brangus cattlemen in the nation.

50. C. W. W. to Samuel D. Myres, March 3, 1965, in the authors' possession.

51. O. W. W. to J. C. W., September 18, 1942, Haley Library.

52. Myres, ed., *Pioneer Surveyor, Frontier Lawyer,* 316; O. W. W. to J. C. W., June 8, 1943, Haley Library; and Herns to Pollard.

53. Ben G. Smith, "Remembrance of Judge O. W. Williams of Fort Stockton," included with Ben G. Smith to C. W. W., September 29, 1964, Haley Library.

54. Pecos County Water Control & Improvement Dist. No. 1 v. Williams, 271 S.W.2d 503 (Tex. Civ. App.–El Paso 1954).

55. Glenn Justice, "Comanche Springs," *Handbook of Texas Online,* http://www.tshaonline.org/; Allan Freedman, "Clayton Williams and West Texas Water," *Texas Observer,* July 27, 1990; and Daggett, *Pecos County History,* 1:288–91.

56. Daggett, *Pecos County History,* 1:288–91; and "Big Cast Works Diligently for Cavalcade Scenes," *Fort Stockton Pioneer,* June 10, 1949.

57. Brief for Appellees, February 1, 1954, 9, 9n; Pecos County Water Control and Improvement District No. 1, Appellant, v. Clayton W. Williams, et al., Appellees, Court of Civil Appeals for the Eighth Supreme Judicial District of Texas at El Paso, Appeal from the Eighty-Third Judicial District Court, Pecos County, Texas.

58. "Pima Cotton," Answers.com, accessed August 25, 2010, http://www.answers.com/. The cotton was named after Pima County, Arizona, the site of the experimental farm on which it was developed.

59. Brief for Appellees, 9; Daggett, *Pecos County History,* 1:145; and Williams, discussions with his family.

60. Ted A. Small and George B. Ozuna, *Ground-Water Conditions in Pecos County, Texas, 1987,* U.S. Geological Survey Water-Resources Investigations Report 92–4190 (Austin: U.S. Department of the Interior, 1993), 31.

61. "Water District Seeking Means of Increasing Flow from Comanche Springs," *Fort Stockton Pioneer,* April 19, 1951; "Comanche Swimming Pool Will Be Completed This Week," *Fort Stockton Pioneer,* October 15, 1953; C. A. Armstrong and L. G. McMillion, *Geology and Ground-Water Resources of Pecos County, Texas* (Austin: Texas Board of Water Engineers, 1961); and Small and Ozuna, 26. In their 1987 report, Small and Ozuna credited a brief resumption of the Comanche Springs flow in October 1986 to "the cessation of irrigation pumpage in August 1986 and several weeks of record or near-record precipitation near Fort Stockton."

62. Williams, discussions with his family.

63. T. C. Railway Co. v. East, 98 Tex. 146, 81 S.W. 279 (1904).

64. *Williams,* 271 S.W. 2d 503.

65. Cochran, *Claytie,* 332.

66. Clay Pollard, e-mail to Louis Gwin; Elmer Kelton, *The Time It Never Rained* (New York: Tom Doherty, 1973); and "North American Drought: A Paleo Perspective," National Climatic Data Center, accessed August 25, 2010, http://www.ncdc.noaa.gov/. Pecos County rainfall totals by month and year are available in the Annual Climatological Summary reports of the NCDC, which were provided to the authors by the Office of the Texas State Climatologist and are online at http://cdo.ncdc.noaa.gov/ancsum/ACS. The figure for the average annual rainfall in Pecos County was taken from data provided by the Fort Stockton weather station and is contained in Wilfried H. Portig, *Atlas of the Climates of Texas: Based on the 50 Year Period 1910–1959* (Austin: Bureau of Engineering Research, University of Texas, 1962), n.p.

67. Scott Pollard, e-mail to Louis Gwin, February 9, 2010.

68. Clay Pollard, tape-recorded reminiscences; and "Leon Springs Pupfish," Nature Conservancy, accessed August 25, 2010, http://www.nature.org/. The endangered pupfish grows to about two inches in length and has disappeared from most of the springs in West Texas, except for the Diamond Y Springs Preserve north of Fort Stockton and a small stretch of Leon Creek west of Fort Stockton.

69. Freedman, "Clayton Williams and West Texas Water," 7.

70. Cochran, *Claytie,* 333; and Russell S. Johnson, "Groundwater Law, Groundwater Planning and Groundwater Management," in *The Changing Face of Water Rights in Texas,* 9th Annual (Austin: State Bar of Texas, 2008), chap. 2.2. Johnson, a specialist in groundwater law for an Austin law firm, wrote in 2008 that "Over one hundred years of jurisprudence have left the [rule of capture] law little changed and much criticized." However, Johnson says that the creation of groundwater conservation districts (or similar government agencies) by the Texas Legislature, and other legislative changes, "have substantially increased the powers of groundwater conservation districts, creating the opportunity for conflict between landowner rights and the exercise of these regulatory powers."

71. "Governing Bodies Need to Come Together to Resolve Current Water Issue," *Midland Reporter-Telegram,* February 6, 2010, http://www.mywesttexas.com/; Sterry Butcher, "Tom Craddock-Clayton Williams Bill: Officials Rally against Proposed Water District," *Big Bend Sentinel,* February 8, 2010, http://www.bigbendsentinel.com/; and "Commissioners Court Opposes Another Water Export Bill," *Fort Stockton Pioneer,* accessed February 8, 2010, http://www.fortstocktonpioneer.com/.

72. Freedman, "Clayton Williams and West Texas Water," 10.

73. "Trial of County Judge Will Reach Jury This Morning," *Fort Stockton Pioneer,* March 19, 1953.

74. Freedman, 10; and George Baker, "Open Letter Of Clayton W. Williams to the County Judge," *Fort Stockton Pioneer,* August 19, 1954. Another issue that Baker believes contributed to Clayton's defeat was a charge that Clayton preferred to hire *braceros* (Mexican laborers) rather than local laborers to harvest crops. Clayton replied that he had tried to hire local Latin Americans but that none were "content to live out on the farm and work steadily."

75. Baker, "Open Letter"; "Correction," *Fort Stockton Pioneer,* August 26, 1954; and Freedman, "Clayton Williams and West Texas Water," 10.

76. C. W. W. to Kathryn Walker, September 2, 1954, in the authors' possession.

77. Clayton W. Williams Jr., discussion with Louis Gwin.

Chapter 11

1. Clayton W. Williams Jr., conversation with Louis Gwin.

2. Ibid.

3. Williams, discussions with his family; and Janet Pollard, discussion with Louis Gwin.

4. Williams, discussions with his family.

5. Clayton W. Williams Jr., discussion with Louis Gwin; and Janet Pollard, discussion with Louis Gwin.

6. "Fort Stockton Man Writes Momumental History of Texas," *Midland Reporter-Telegram,* October 29, 1969; and Clayton W. Williams, "Fort Stockton's First 100 Years," Official Program, *Fort Stockton Centennial, 1859–1959,* June 21–27, 1959.

7. Janet Pollard, discussion with Louis Gwin; Haley et al. to Hall of Fame Committee; and Daggett, "Clayton Williams Sr.," *Pecos County History,* 2:515–16.

8. C. W. W. to Kathryn Walker, January 21, 1954 and January 16, 1955, in the authors' possession.

9. C. W. W. to Ben G. Smith, October 3, 1964, in the authors' possession.

10. C. W. W. to S. D. Myres, December 16, 1965, in the authors' possession.

11. C. W. W. to Carl Hertzog, April 26, 1966, in the authors' possession.

12. C. W. W. to S. D. Myres, April 15, 1966, in the authors' possession.

13. Virginia Holliman Van Horn, "Naylor, Joe Oliver," *Handbook of Texas Online,* http://www.tshaonline.org/. Clayton's explanation for the book's title was found in a newspaper clipping in the Pollard files; the source and date are unknown.

14. Odie B. Faulk, review of *Never Again,* by Clayton Williams, in *Southwestern Historical Quarterly* 73, no. 4 (April 1970): 561–62.

15. Linda Anderson, "Area Author and Historian Has 'Interesting Life,'" *Midland Reporter-Telegram,* June 13, 1982.

16. Janet Pollard, conversation with Louis Gwin; and "Stockton Author Celebrates Today," *San Angelo Standard-Times,* October 11, 1969.

17. Clay Pollard, e-mail message to Louis Gwin.

18. Williams, discussions with his family.

19. Janet Pollard, discussion with Louis Gwin.

20. C. W. W. to Smith.

21. Clayton W. Williams Jr., discussion with Louis Gwin.

22. C. W. W. to Frank T. Pickrell, August 6, 1969, in the authors' possession; C. W. W. to Berte Haigh, August 12, 1969, in the authors' possession; and Berte Haigh, *Land, Oil and Education* (El Paso: Texas Western, 1986).

23. Clay Pollard, tape-recorded reminiscences. The photograph of Clay Pollard at the signing was published in the *Fort Stockton Pioneer,* December 12, 1974.

24. W. Frank Blair, review of *Animal Tales,* in *West Texas Historical Association Yearbook* 50 (1974): 177–78. For Clayton's views on predators, see *Animal Tales,* xi, 221.

25. Clay Pollard, tape-recorded reminiscences.

26. Clayton published "Excerpts from the Diary of George Wedemeyer," *West Texas Historical Association Yearbook* 46 (1970): 156–66; "The Howard's Well Massacre, 1872," *West Texas Historical Association Yearbook* 51 (1975): 58–62; "A Threatened Mutiny of Soldiers at Fort Stockton in 1873 Resulted in Penitentiary Sentences," *West Texas Historical Association Yearbook* 52 (1976): 78–83; "The Topographical Ghost—Horsehead Crossing!" *Permian Historical Annual* 17 (1977): 37–56; and "The Pontoon Bridge of the Pecos, 1869–1886," *Permian Historical Annual* 24 (1978): 51–60. He also wrote several book reviews for the *West Texas Historical Association Yearbook.*

27. Williams, "Topographical Ghost," 37–56; Dalton King, "Horsehead Crossing on the Rio Pecos," in Daggett, *Pecos County History,* 1:68–70; and Glenn Justice, "Horsehead Crossing," *Handbook of Texas Online,* http://www.tshaonline.org/.

28. Clayton W. Williams Jr., discussion with Louis Gwin.

29. Ibid.

30. C. W. W. to Mari Helen Schultz, August 28, 1960, in the authors' possession.

31. Janet Pollard, discussion with Louis Gwin.

32. Clayton W. Williams Jr., discussion with Louis Gwin.

33. Clay Pollard, e-mail message to Louis Gwin.

34. Janet Pollard and Scott Pollard, discussion with Clay Pollard; and Cochran, *Claytie,* 16–17.

35. Janet Pollard and Scott Pollard, discussion with Clay Pollard.

36. Ibid.

37. Clay Pollard, e-mail message to Louis Gwin.

38. Janet Pollard, discussion with Louis Gwin.

39. Janet Pollard and Scott Pollard, discussion with Clay Pollard; and Clay Pollard, Scott Pollard, and Janet Pollard, discussion with Louis Gwin.

40. Janet Pollard, discussion with Louis Gwin.

41. Janet Pollard, telephone interview with Louis Gwin, October 6, 2009.

42. Scott Pollard, e-mail message to Louis Gwin, August 2, 2009.

43. Williams, *Texas' Last Frontier,* xiv.

44. Brad Agnew, review of *Texas' Last Frontier,* in *Southwestern Historical Quarterly* 87 (1983–84): 108–109.

45. Herbert H. Lang, "Texas A&M University Press Publishes 'First Rate History of Lasting Significance,'" *Texas Aggie,* March 1982, 6.

46. Judith Parsons, review of *Texas' Last Frontier,* in *The West Texas Historical Association Year Book* 54 (1983): 206–208.

47. Janet Pollard and Scott Pollard, discussion with Clay Pollard.

48. Janet Pollard, written reminiscence, in the authors' possession; and Herns to Pollard.

49. Janet Pollard, written reminiscence.

50. Cochran, *Claytie,* 116–27.

51. Janet Pollard, discussion with Louis Gwin.

52. Cochran, *Claytie,* 23.

53. Janet Pollard, written reminiscence.

54. Haley et al. to Hall of Fame Committee. Clayton served one term as commander of the Fort Stockton VFW and twice as commander of the local American Legion post, where he was a member for more than sixty years.

55. Clayton W. Williams Jr., discussion with Louis Gwin.

56. Clay Pollard, tape-recorded reminiscence.

57. Janet Pollard and Scott Pollard, discussion with Clay Pollard.

58. Clay Pollard, tape-recorded reminiscence.

Epilogue

1. Frank Baker, "Pipelines," *Fort Stockton Pioneer,* September 15, 1983.

2. Ibid.

Bibliography

Agnew, Brad. Review of *Texas' Last Frontier: Fort Stockton and the Trans-Pecos, 1861–1895,* by Clayton Williams. *Southwestern Historical Quarterly* 87 (1983–1984): 108–109.

Anderson, Robert O. *Fundamentals of Petroleum Geology.* Norman: University of Oklahoma Press, 1984.

Armstrong, C. A., and L. G. McMillion. *Geology and Ground-Water Resources of Pecos County, Texas.* Texas Board of Water Engineers Bulletin 6106, vol. 1, no. 45. Austin: Texas Board of Water Engineers, 1961.

Arthur, Robert. *The Coast Artillery School, 1824–1927.* Fort Monroe, Virginia: Coast Artillery School Press, 1928.

Barry, Louise. "Kansas before 1854: A Revised Annals." *Kansas Historical Quarterly* 28, no. 1 (1962): 25–59.

Batchelor, John, and Ian Hogg. *Artillery.* New York: Charles Scribner's Sons, 1972.

Blair, W. Frank. Review of *Animal Tales of the West,* by Clayton Williams. *West Texas Historical Association Yearbook* 50 (1974): 177–78.

Brown, Meredith Mason. *Frontiersman: Daniel Boone and the Making of America.* Baton Rouge: Louisiana State University Press, 2008.

Bruce, Robert B. *A Fraternity of Arms: America and France in the Great War.* Lawrence: University of Kansas Press, 2003.

"Degrees and Certificates Conferred at the Forty-Eighth Annual Commencement." *Bulletin of the Agricultural and Mechanical College of Texas* 11, no. 4 (1925): 323–24.

Caidin, Martin. *Barnstorming.* New York: Duell, Sloan and Pearce, 1965.

Carter, Harvey Lewis. *The Life and Times of Little Turtle: First Sagamore of the Wabash.* Urbana: University of Illinois Press, 1987.

Chandler, C. DeF. "Military Observation Balloons." *Field Artillery Journal* 7, no. 1 (1917): 15–20. http://sill-www.army.mil/FAMAG/archives.htm.

Cochran, Mike. *Claytie: The Roller-Coaster Life of a Texas Wildcatter.* College Station: Texas A&M University Press, 2007.

Conolly, W. R. "Motor Transportation for Artillery." *Field Artillery Journal* 9, no. 3 (1919): 255–75.

Coulter, Charles C. "Petroleum 'Graveyard Area' of Texas." *Oil and Gas Journal* 17, no. 48 (1919): 66–67.

Covington, James W., ed. "A Robbery on the Santa Fe Trail, 1827." *Kansas Historical Quarterly* 21, no. 7 (1955): 560–63.

Cox, G. H., C. L. Dake, and G. A. Muilenburg. *Field Methods in Petroleum Geology.* New York: McGraw Hill, 1921.

Crabtree, James D. *On Air Defense.* Westport, Connecticut: Praeger, 1994. http://books.google.com/.

Daggett, Marsha Lea, ed. *Pecos County History.* 2 vols. Canyon, Texas: Staked Plains, 1984.

Darrow, Margaret H. *French Women and the First World War: War Stories of the Home Front.* Oxford: Berg, 2000.

DeArment, Robert K. *George Scarborough: The Life and Death of a Lawman on the Closing Frontier.* Norman: University of Oklahoma Press, 1996.

Dethloff, Henry C. *A Pictorial History of Texas A&M University, 1876–1976.* College Station: Texas A&M University Press, 1975.

Dickey, Philip S., III. *The Liberty Engine, 1918–1942.* Smithsonian Annals of Flight, vol. 1, no. 3. Washington: Smithsonian Institution Press, 1968.

Dittmar, Gus C. *They Were First: Recollections of the First Officers Training Camp of Leon Springs, Texas, May 8 to August 15, 1917.* Austin: Steck-Warlick, 1969.

Dock, Lavinia L., Sarah Elizabeth Pickett, Clara D. Noyes, Fannie F. Clement, Elizabeth G. Fox, and Anna R. Van Meter. *History of American Red Cross Nursing.* New York: MacMillan, 1922. http://books.google.com/.

Edel, Wilbur. *Kekionga! The Worst Defeat in the History of the U.S. Army.* Westport, Connecticut: Praeger, 1997.

Ellis, John, and Michael Cox. *The World War I Databook.* London: Aurum, 2001.

Farwell, Byron. *The United States in the Great War, 1917–1918.* New York: Norton, 1999.

Faulk, Odie B. Review of *Never Again,* by Clayton Williams. *Southwestern Historical Quarterly* 73, no. 4 (April 1970): 561–62.

Goldberg, Ronald Allen. *America in the Twenties.* Syracuse, New York: Syracuse University Press, 2003.

Hager, Dorsey. *Practical Oil Geology.* New York: McGraw Hill, 1916.

Haigh, Berte. *Land, Oil and Education.* El Paso: Texas Western Press, 1986.

Hamburger, Kenneth E. *Learning Lessons in the American Expeditionary*

Forces. CMH Pub 24–1. [Washington, D.C.?]: U.S. Army Center of Military History, 1997.

Hart, A. E. "The Callaway Family of Virginia." Unpublished manuscript, 1929. Authors' collection.

———. "Col. Richard Callaway and Allied Families." 2 vols. Unpublished manuscript, 1935. Authors' collection.

Hempson, J. G. G. "The Automobile Engine, 1920–1950." In *A History of the Automotive Internal Combustion Engine.* Warrendale, Pennsylvania: Society of Automotive Engineers, 1976.

Hogg, Ian. *Allied Artillery of World War One.* Wiltshire, United Kingdom: Crowood, 1998.

Hotz, Robert, ed. *Way of a Fighter: The Memoirs of Claire Lee Chennault.* New York: C. P. Putnam's Sons, 1949.

James, Bill C., and Mary Kay Shannon. *Sheriff A. J. Royal: Fort Stockton, Texas.* Np: B. C. James and M. K. Shannon, 1984.

Johnson, Russell S. "Groundwater Law, Groundwater Planning and Groundwater Management." In *The Changing Face of Water Rights,* 9th Annual. Austin: State Bar of Texas, 2008.

Katz, Friedrich. *The Life and Times of Pancho Villa.* Stanford: Stanford University Press, 1998.

Keegan, John. *The First World War.* New York: Vintage, 2000.

Keene, Jennifer D. *Doughboys, the Great War, and the Remaking of America.* Baltimore: Johns Hopkins University Press, 2001.

Keiger, J. F. V. *Raymond Poincaré.* Cambridge: Cambridge University Press, 1997.

Kelton, Elmer. *The Time It Never Rained.* New York: Tom Doherty, 1973.

Lang, Herbert H. "Texas A&M University Press Publishes 'First Rate History of Lasting Significance.'" Review of *Texas' Last Frontier: Fort Stockton and the Trans-Pecos, 1861–1895,* by Clayton Williams. *Texas Aggie,* March 1982, 6.

Langenkamp, R.D. *Handbook of Oil Industry Terms & Phrases.* 4th ed. Tulsa: PennWell, 1984.

The Long Horn. College Station: Texas Agricultural and Mechanical College, 1915.

Mac Donald, Laura M. *Curse of the Narrows.* New York: Walker, 2005.

Mead, Frederick S., ed. *Harvard's Military Record in the World War.* Boston: Harvard Alumni Association, 1921. http://books.google.com/.

Moody, John. *Moody's Analyses of Investments.* Part II, *Industrial Investments.* New York: Moody's Investors Service, 1920.

Moody's Analyses of Investments and Security Rating Books, Industrial Investments. New York: Moody's Investors Service, 1923.

Morgenthaler, George. *The River Has Never Divided Us: A Border History of La Junta de los Rios*. Austin: University of Texas Press, 2004.

Myres, Samuel D. *The Permian Basin: Petroleum Empire of the Southwest*. Vol. 1, *Era of Discovery, from the Beginning to the Depression*. El Paso: Permian Press, 1973.

————. *The Permian Basin: Petroleum Empire of the Southwest*. Vol. 2, *Era of Advancement, from the Depression to the Present*. El Paso: Permian Press, 1977.

Olien, Roger M., and Diana Davids Hinton. *Wildcatters: Texas Independent Oilmen*. College Station: Texas A&M University Press, 2007.

Ottosen, P. H., ed. *Trench Artillery A.E.F.: The Personal Experiences of Lieutenants and Captains of Artillery Who Served with Trench Mortars*. Boston: Lothrop, Lee & Shepard, 1931.

Owen, Edgar Wesley. *Trek of the Oil Finders: A History of Exploration for Petroleum*. Tulsa: American Association of Petroleum Geologists, 1975.

Parsons, Judith. Review of *Texas' Last Frontier: Fort Stockton and the Trans-Pecos, 1861–1895,* by Clayton Williams. *The West Texas Historical Association Year Book* 54 (1983): 206–208.

"Past Year Was Busiest in the History of Petroleum." *Oil and Gas Journal* 18, no. 34 (1920): 54.

Payne, Walter A., ed. *Benjamin Holt: The Story of the Caterpillar Tractor*. Stockton, California: University of the Pacific, 1982.

Perrett, Geoffrey. *America in the Twenties: A History*. New York: Simon and Schuster, 1982.

Pope, Clarence. *An Oil Scout in the Permian Basin, 1924–1960*. El Paso: Permian, 1972.

Pope, Stephen, and Elizabeth-Anne Wheal. *The Dictionary of the First World War*. New York: St. Martin's, 1995.

Portig, Wilfried H. *Atlas of the Climates of Texas: Based on the 50 Year Period 1910–1959*. Austin: Bureau of Engineering Research, University of Texas, 1962.

Prindle, David F. "Oil and the Permanent University Fund: The Early Years." *Southwestern Historical Quarterly* 86, no. 2 (October 1982): 277–98.

Quaife, Milo Milton. *Kit Carson's Autobiography*. Lincoln: University of Nebraska Press, 1966.

Railroad Commission of Texas. *Oil and Gas Division Annual Report, 2006*. Austin: Railroad Commission of Texas, 2007.

Sanderson, Susan R. Walsh. *Land Reform in Mexico, 1910–1980*. Orlando: Academic Press, 1984.

Schwettmann, Martin W. *Santa Rita.* Austin: Texas State Historical Association, 1943.

Sides, Hampton. *Blood and Thunder: An Epic of the American West.* New York: Doubleday, 2006.

Siler, Joseph F. *The Medical Department of the United States Army in the World War.* Vol. 9, *Communicable and Other Diseases.* Washington, DC: U.S. Government Printing Office, 1928.

Small, Ted A., and George B. Ozuna. *Ground-Water Conditions in Pecos County, Texas, 1987.* U.S. Geological Survey Water-Resources Investigations Report 92–4190. Austin: U.S. Department of the Interior, 1993.

Sterling, Bryan B., and Frances N. Sterling. *Will Rogers and Wiley Post: Death at Barrow.* New York: M. Evans, 1993.

Synder, Alice Ziska, and Milton Valentine Snyder. *Paris Days and London Nights.* New York: E. P. Dutton, 1921. http://books.google.com/.

Taylor, John W. R., ed. *Jane's Fighting Aircraft of World War I.* London: Random House, 2001.

Thévenot, Roger. *A History of Refrigeration throughout the World.* Translated by J. C. Fidler. Paris: International Institute of Refrigeration, 1973.

Townsend, Mary Alice Happle. "Andrew Jackson Royal." *Permian Historical Annual* 24 (1985): 51–60.

Travers, Tim. *How the War Was Won: Command and Technology in the British Army on the Western Front, 1917–1918.* London: Routledge, 1992.

Venzon, Anne Cipriano, ed. *The United States in the First World War: An Encyclopedia.* New York: Garland, 1995.

Votaw, John F. *The American Expeditionary Forces in World War I.* Oxford: Osprey, 2005.

War Department. *Manual for Trench Artillery, United States Army (Provisional).* Part 1, *Trench Artillery.* [Washington,DC?]: Office of the Adjutant General, 1918. Reprint, 2006. http://cgsc.cdmhost.com/u?/p4013co119, 133.

Wecter, Dixon. *When Johnny Comes Marching Home.* Cambridge: Riverside, 1944.

Williams, Clayton Wheat. *Animal Tales of the West.* San Antonio: Naylor, 1974.

———. "Excerpts from the Diary of George Wedemeyer." *West Texas Historical Association Yearbook* 46 (1970): 156–66.

———. "Fort Stockton's First 100 Years." In *Fort Stockton Centennial, 1859–1959.*

———. "The Howard's Well Massacre—1872." *West Texas Historical Association Yearbook* 51 (1975): 58–62.

————. *Never Again.* 3 vols. San Antonio: Naylor, 1969.

————. Papers. Haley Memorial Library and History Center, Midland, Texas.

————. "The Pontoon Bridge of the Pecos, 1869–1886." *Permian Historical Annual* 24 (1978): 51–60.

————. *Texas' Last Frontier: Fort Stockton and the Trans-Pecos, 1861–1895.* College Station: Texas A&M University Press, 1982.

————. "A Threatened Mutiny of Soldiers at Fort Stockton in 1873 Resulted in Penitentiary Sentences." *West Texas Historical Association Yearbook* 52 (1976): 78–83.

————. "The Topographical Ghost—Horsehead Crossing!" *Permian Historical Annual* 17 (1977): 37–56.

Williams, Ermine. Unpublished manuscript, n.d. Fort Stockton Public Library.

Williams, Oscar Waldo. Papers. Haley Memorial Library and History Center, Midland, Texas.

————. Papers, 1894–1943. Center for American History, University of Texas at Austin.

————. *Pioneer Surveyor, Frontier Lawyer: The Personal Narrative of O. W. Williams, 1877–1902.* Edited and annotated by S. D. Myres. With an introduction by C. L. Sonnischen. El Paso: Texas Western College Press, 1966.

Willis, Michael. *Billy the Kid: The Endless Ride.* New York: Norton, 2007.

Zieger, Robert H. *America's Great War: World War I and the American Experience.* Lanham, Maryland: Rowman & Littlefield, 2000.

Index

Page numbers in *italics* refer to illustrations.